Defrosting Ancient Microbes

Defrosting Ancient Microbes

Emerging Genomes in a Warmer World

Scott O. Rogers
Professor of Molecular Biology and Evolution
Bowling Green State University

John D. Castello
Professor Emeritus of Microbiology and Forest Pathology
SUNY College of Environmental Science and Forestry

CRC Press
Taylor & Francis Group
Boca Raton London New York

CRC Press is an imprint of the
Taylor & Francis Group, an **informa** business

CRC Press
Taylor & Francis Group
6000 Broken Sound Parkway NW, Suite 300
Boca Raton, FL 33487–2742

© 2020 by Taylor & Francis Group, LLC

CRC Press is an imprint of Taylor & Francis Group, an Informa business

No claim to original U.S. Government works

International Standard Book Number-13: 978-0-367-22350-2 (Hardback)
International Standard Book Number-13: 978-0-367-22262-8 (Paperback)

A catalog record for this book has been requested

**Visit the Taylor & Francis Web site at
www.taylorandfrancis.com**

**and the CRC Press Web site at
www.crcpress.com**

Contents

Preface

Mix 99 parts water with 1 part microbes, stir, and freeze. What you get is frozen microbes that will be alive and well long after you, your children, your grandchildren, and most of your descendants are gone. Some of the microbes may live on (if kept frozen) for millions of years. Some of the life on Earth has lived through a multitude of catastrophic events on Earth. Events that would have killed most, if not all, of the humans on Earth. Scientists studying ancient ice can measure the amounts of gases that were in the atmosphere, and were entrapped in the ice, hundreds of thousands of years ago. They can measure the temperature of the Earth during those times, as well. Other researchers can examine the microbes, and even larger organisms, that were alive at that time, and reconstruct what parts of the Earth looked like during those times. It is as if recording instruments had existed all over the world during all of human existence, and today, we are able to read those instruments. Ice on Earth, and even on other planets, holds these records, and sometimes it holds living organisms that can be revived and studied.

We started writing the initial portions of this book in 2000, at a time when investigating microbes in ancient ice was in its infancy, and there was very little information to report. At that time, the field of Astrobiology was moving from a small area of research into a well-funded major branch of science. Also, global climate change had been firmly established in the scientific community, but still was debated by politicians and the public. The result of the preliminary nature of all of this resulted in a book manuscript that was thin on data and heavy on speculation. About three years ago, we returned to the book manuscript and established that a great deal of data had been produced over the past two decades, such that it would be possible to produce a stronger book. This was not only caused by the increase in data, but also by the increase in the knowledge of the public in these areas. Many popular press articles have been written, and many programs have been produced discussing ancient organisms, the many aspects of exploring planets and moons in the Solar System, and the big changes that have occurred in greenhouse gases, melting, ice, sea level rise, and the human causes of these. In this book, we detail the discoveries and conclusions in the research on ancient ice, and tie it in with the discoveries in astrobiology, epidemiology, global climate changes, and sea level rises.

While ice contains a valuable record, it also encases microbes that are beneficial (e.g., those that recycle nutrients, those that fix carbon from carbon dioxide, those that help plants to soak up water and nutrients, those that produce antibiotics, and a multitude of other vital processes), as well as those that are dangerous. There are many pathogenic microbes encased in environmental ice around the world. All are released as glaciers melt. With the increases in melting of ice worldwide, more of these organisms are being released. The potential for disease outbreaks, epidemics, and pandemics is real, as is the loss of beneficial microbes. Disease outbreaks have likely occurred before from glacial melting, but it is unknown when and how frequently.

Global climate change threatens the loss of this extremely valuable resource, and at the same time, potentially exposes the world to pathogens. Studies of ancient ice compared to real time measurements of atmospheric gases taken over the past century have led to the realization that humans are in the process of changing the Earth in ways that are causing changes in the land, ocean, and atmospheric composition and temperature, as well as altering precipitation, wind, storms, ocean currents, and melting ice worldwide, which is causing sea levels to rise, and increased heating due to the loss of the reflective nature of snow and ice. As mentioned above, this also is causing unprecedented releases of microbes, including pathogens around the world. Scientists and non-scientists alike are worried. Hopefully, it will not be a deadly pandemic that finally leads people and governments to act decisively to slow or reverse these global changes. With increases in world temperatures, melting of ice, and human population increase, we are approaching a time when the risks of a negative outcome in the near future are growing at an alarming rate.

The science of Astrobiology involves the search for extraterrestrial life. This is directly related to research of microbes in environmental ice and permafrost (frozen soils). During the past half century, a great deal has been learned about the planets and moons in the Solar System and beyond. At one point in time, our vision was limited to telescopes that could help us see the Moon, Mercury, Venus, Mars, and blurry visions of Jupiter, Saturn, Uranus, Neptune, and Pluto. At that time, it looked like Earth was the only planet that could sustain life. No other planet or moon had enough water and the right temperatures to harbor life. Today, we know that this view was wrong. There is a lot of water in our Solar System, although much of it is in the form of ice. But, large lakes and oceans also exist. By studying life in ice on Earth, we can reach conclusions about which planets and moons might have life on them. This provides a guide to help plan missions and probes to search for life in intelligent ways.

During our work with ice cores, we met many researchers. While we cannot name all of them, they were all interesting and helpful, including those from France, Russia, Belgium, Denmark, Canada, Tasmania, the UK, and the US. We also met engineers, electricians, welders, carpenters, drillers, and NSF officials. They should not be forgotten in work obtaining the ice cores. They were just as dedicated, interesting, and helpful as were the researchers. Many students also were involved, and were vitally important to the research endeavors and results. It really does take a community to do this complex, challenging, and often dangerous work.

This book is arranged into 20 chapters. Chapter 1 is an overview of working with ice, and isolating and studying the organisms encased in the ice. Chapter 2 deals primarily with the overall implications of the work, including the importance to studies of evolution, epidemics, applied projects (e.g., antibiotics, food products), and the search for life away from the Earth. Chapter 3 presents a discussion of the properties of water and ice, and why ice is such a great preservative of organisms. Chapter 4 explores the diversity of environmental ice, describing similarities and differences between glaciers, ice fields, ice domes, sea ice, and lake ice. Chapter 5 is all about permafrost, which is simply frozen soil, but as with environmental ice, there are many different types of permafrost. Chapter 6 discusses the definitions of life, and how to identify living and dead microbes. Chapter 7 defines and discusses fossils, both dead and living. Chapter 8 explains some of the steps in the planning, funding, drilling, and analyzing inclusions in ice cores. Chapter 9 outlines some of the major methods for isolating and identifying organisms entrapped in the ice. Chapter 10 outlines the history of the studies of microbes in ice, including the confluence of genetics, evolutionary biology, geology, ice biology, molecular biology, and microbiology. Each of these had to exist for the study of microbes in ice to be possible. Chapter 11 describes the details of what has been found in the ice during the past 60 years. Chapter 12 is a detailed account of what has been found in subglacial Lake Vostok, which has been covered for 15 million years by thousands of meters of glacial ice. The short story is that it contains a diversity of organisms. Chapter 13 discusses other lakes, including subglacial Lake Whillans, as well as ice from Lake Erie and surface lakes in Antarctica. Chapter 14 discusses the types of living organisms that have been extracted from ancient ice. The bottom line is that they are very adaptable to a broad number of conditions, including living through freeze-thaw cycles. Chapter 15 describes the hazardous organisms that have been found in ice, and that are capable of causing disease in animals (including humans), plants, and other organisms. This chapter primarily deals with viruses. Chapter 16 discusses the hazardous bacteria and other microbes that have been found in ice, or that are capable of surviving freezing and thawing. It also discusses the dangers to people who work with the ice. It probably is not a good idea to put a glacial ice cube into your drink. Chapter 17 discusses the process of genome recycling, which is the process whereby an organism is frozen for long periods of time, after which it melts out of the ice and enters a population (or a host) carrying with it the ancient genome that now enters the more recent genome. Chapter 18 is a discussion of where in the Solar System to look for signs of life. There are quite a few planets and moons that have the potential to harbor life. Chapter 19 discusses global climate change, and how it affects the study of microbes in ice, as well as many other aspects of life on Earth. Chapter 20 wraps up and summarizes the major concepts and findings that have come from the study of ice and the microbes within ice.

Acknowledgments

I (SOR) thank my wife, Mary, son, Ben, and daughter, Liz, for moral support. I could not have written the book without their support and understanding. I also thank my mother, father, sister, and brother for their encouragement throughout the years. I thank all my students and colleagues (whether named in the book or not) who helped me in the many projects, and with whom I discussed many of the topics included in this book. Special thanks to Chuck Crumly at Taylor & Francis for his help in publishing this book. I also thank colleagues at BGSU for their kind words about the book, and to BGSU for allowing me the time to complete the writing of this book. JDC wishes to thank his graduate students for conducting portions of the virus research cited in this book.

Authors

Scott O. Rogers is a professor of molecular biology and evolution in the Department of Biological Sciences at Bowling Green State University, Bowling Green, Ohio. He received his BS (1976) and MS (1980) degrees in Biology from the University of Oregon, Eugene, Oregon; and a PhD (1987) in Plant Molecular Biology from the University of Washington, Seattle, Washington; and was then a postdoc for two years at the same university. He was an Assistant Professor and Associate Professor at the State University of New York, College of Environmental Science and Forestry, Syracuse, New York from 1989 through 2001, before moving to BGSU, as Professor (and Departmental Chair 2001–2011). He has taught courses in general biology, botany, cell biology, molecular biology, molecular genetics, molecular techniques, molecular evolution, and bioinformatics. Research in his lab includes studies of microbes and nucleic acids preserved in ice, life in extreme environments, group I introns, ribosomal RNA genes, ribosomes, evolution of the genetic code, molecular microbial phylogenetics, microbial metagenomics/metatranscriptomics, ancient DNA, and plant development.

John D. Castello is professor emeritus of microbiology and forest pathology in the Department of Environmental and Forest Biology, State University of New York, College of Environmental Science and Forestry, Syracuse, New York. He received his BA (1973) in Biology from Montclair State College, Upper Montclair, New Jersey; his MS (1976) in Plant Pathology from Washington State University, Pullman, Washington; and his PhD (1978) in Plant Pathology from the University of Wisconsin, Madison, Wisconsin. He has been Assistant, Associate, Full Professor, and Associate Chair at SUNY-ESF, Department of Environmental and Forest Biology, Syracuse, New York. He has taught courses in microbiology, forest pathology, plant virology, forest health, and peoples, plagues, and pests.

List of Figures

List of Tables

Abbreviations

bybp	billion years before present
cDNA	complementary DNA (copy of RNA)
CE	common era or current era
CH_4	methane
cm	centimeter(s)
CO_2	carbon dioxide
DNA	deoxyribonucleic acid
EDTA	Ethylendiaminetetraacetic acid
EPICA	European Project for Ice Coring in Antarctica
ft	foot (feet)
GISP	Greenland Ice Sheet Project
GRIP	Greenland Ice Core Project
H1N1	influenza A strain H1N1 (haemagglutinin 1, neuraminidase 1)
IGY	International Geophysical Year (1957/58)
km	kilometer(s)
km^2	square kilometer(s)
km^3	cubic kilometer(s)
l	liter(s)
m	meter(s)
Mg^{2+}	magnesium ion
mi	mile(s)
mi^2	square mile(s)
mi^3	cubic mile(s)
ml	milliliter(s)
µl	microliter(s)
mybp	million years before present
NEEM	North Greenland Eemian Ice Drilling Project
NICL	National Ice Core Laboratory
NSF-ICL	National Science Foundation Ice Core Laboratory
NGRIP	North Greenland Ice Core Project
NGS	next generation sequencing
PBSY	phage-like element in *Bacillus subtilis* Y
PCR	polymerase chain reaction
pH	negative log of proton (H^+) concentration, low pH is acidic, pH 7 is neutral, high pH is alkaline
RNA	ribonucleic acid
SEM	scanning electron microscope (microscopy)
TEM	transmission electron microscope (microscopy)
TMV	tobacco mosaic tobamovirus

ToMV	tomato mosaic tobamovirus
USSR	Union of Soviet Socialist Republics
VESV	vesicular exanthema swine virus
WAIS	West Antarctic Ice Sheet (WAIS Divide ice core project)
ybp	years before present

1 Reaching Backwards

"Life can only be understood backwards; but it must be lived forwards"

Søren Kierkegaard

REVIVING THE "DEAD"

Can the dead be revived? In Sir Arthur Conan Doyle's "The Lost World," a professor discovers an ancient world suspended in time. Life is as it was millions of years ago. It is a world in which "the ordinary laws of nature are suspended. The various checks which influence the struggle for existence in the world at large are all neutralized." This was science fiction when it was published in 1912. But it is not so far-fetched. What if there were a way to reach back in time and drag ancient life into the present? Actually, it has been occurring naturally on Earth for millions, if not billions of years. Scientists have recently done the same.

Prehistoric organisms are living and growing today. We can revive them, or they can spontaneously revive by melting out of their icy tombs. They are probably on your skin and in your hair at this very moment. Some are shades of russet, bronze, and mahogany. Others have soft, mist-like filaments reaching into the air. Some look like microscopic space ships. There are those that appear like an opaque slime with a gelatinous glow. And others are so very tiny that, in one quick breath, hundreds can effortlessly be drawn into your lungs and into your blood. Quite a long and marvelous journey for a prehistoric spoor.

You might think this is unbelievable, but there are special places on Earth where ancient life is preserved. While this may generate visions of crystallized amber or the tombs of Egypt, there is a much more common and effective preservation method for sustaining life, common on Earth, and utilized every day by you and everyone you know. It is ice. Humans have used ice for centuries to preserve foods, but it has been a natural preservative of life for millions of years. Much evidence has accumulated over the past 200 years, primarily in the past 30 years, indicating that many microbes, as well as some multicellular organisms, can be frozen and then revived when the ice melts decades, centuries, or millennia later. Yearly, millions of tons of organisms are entrapped in ice, and currently many more are melting out of the ice due to global climate change, and the warming that it has caused.

In the polar regions, much of the ice remains frozen, sometimes for millions of years. It continually builds up millennium after millennium, eventually melting or subliming, thus releasing its cargo of viable microorganisms. Ice has formed for so long that excavating deep within its recesses one can construct a timeline long into the past. Just as sedimentary rock layers have been laid down and preserve different sequential points of time, so too are layers of ice laid down over time to preserve a sequential record of past times. In layer after layer, the pages of history can be retraced, back to 1930, 1800, 1200, 500; and to 1000 ybp (years before present), 5000 ybp, 100,000 ybp, 400,000 ybp, 2 million ybp, and beyond; including times beyond human existence (Fig. 1.1). Times of global warming, cooling, volcanic eruptions, disease epidemics, pandemics, and other events can be investigated in these moments frozen in time within ice. Some of the ice buried deep beneath the surface of the Earth is so old that it is difficult to imagine that time and place. It is certainly out of our day-to-day experience. There are glaciers and ice domes that are miles deep in Greenland and Antarctica from which ice cores have been extracted (Fig. 1.2). Other slices through time exist on the surface, having been pushed up and out by the tremendous forces of the glaciers. Sections of this ice provide rich images of the distant past. But this is much more than an academic curiosity. Prehistoric life can actually be incorporated into our modern world. A virus coughed up from a Neanderthal's lungs might emerge

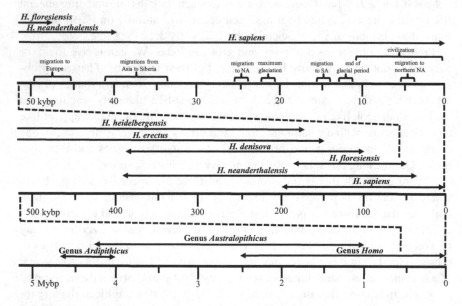

FIGURE 1.1 Timelines of the past 5 million years. This spans the ages of ice cores from Earth currently available for scientific study, as well as the span of several species in the genus *Homo*. At the top is a scale representing the past 50,000 years, indicating the species of *Homo* present, the major human migrations (often coinciding with global climate changes), the last major ice age, and the extent of human civilization (approx. 11,000 years). The middle timeline shows the extent of members of the genus *Homo* during the past 500,000 years. *Homo sapiens* have existed for approximately the last 200,000 years. The lower timeline shows the spans for three genera of human-like species, including *Ardipithicus*, *Australopithicus*, and *Homo*.

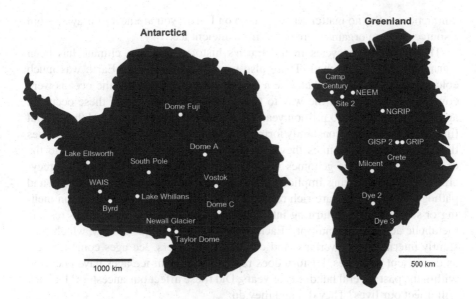

FIGURE 1.2 Maps of Antarctica and Greenland, showing the locations of some of the major ice cores that have been analyzed by scientists. The cores from Greenland (GISP 2, Greenland Ice Sheet Project; and others) and Antarctica (Vostok, Dome C, and others), have yielded information on CO_2 and methane levels, as well as on temperature and microorganisms, reaching back as far as 800,000 ybp.

today, having been in suspended animation for 50,000 years. An influenza or smallpox virus strain that has been frozen for decades, centuries, or millennia may melt from a glacier today to infect a population of humans never exposed to these specific virus strains. Thus, there would likely be no remnant immunity in this population. These immunologically naive populations would be extremely vulnerable to these completely unknown pathogenic viral strains.

Glaciers and polar ice caps are constantly forming and melting. Global warming events accelerate the rate of release of these frozen organisms. During ice ages more of the organisms are entrapped in glaciers. Most of the terrestrial freshwater ice on Earth is in Antarctica (over 91%) and in the Arctic (about 8%), and this is approximately 70% of all of the freshwater on Earth. But there also are glaciers in the Alps, the Southern Alps, the Rockies, the Andes, the Himalayas, and many other locales, including on Mt. Kirinyaga (formerly Mt. Kenya) and Mt. Kilimanjaro near the equator in Africa (although the ice on these two mountains has almost completely disappeared during the past two decades). The water from these glaciers ends up in streams, lakes, and oceans. Many people drink the water and eat the plants and animals that use these sources of water, not realizing that they may be ingesting organisms that are thousands of years old, or older. Microbes are transported to the glaciers by wind, rain, snow, fog, or by macroorganisms. Some of the transport is over very long distances. For example, biological specimens originating from close to the equator have been found in Arctic and

Antarctic ice. So, no matter where you go on Earth, you are never far away from a source of these organisms released from ancient ice.

There have been cycles in the Earth's history when the climate has been warmer and the ice melted. There also were cycles when the Earth was much colder and might have looked like a huge snowball when all of the oceans were covered by thick ice all the way to the equator. The last one of these occurred approximately 600–700 million years ago. Ice has waxed and waned throughout time. Indeed, glaciers constantly form at one end and melt at the other. Sometimes they recede and sometimes they advance. Ice constantly cycles and so do the organisms and unique genomes within them. This organism and genome recycling has broad-ranging implications in terms of evolution, epidemiology, and pathogenicity. Glaciers are rich repositories of entrapped organisms. Upon melting or sublimation (ice turning into water vapor), some of the microbes resume metabolic activity and multiply. Each organism has a unique history, which constantly intermingles, overlaps, and collides with others. Ice ages come and go, as do ancient microbes. History does repeat itself. Mini ice ages have occurred within the past several hundreds of years. Did these affect our ancestors? Do they still affect our lives? They did and they do.

HOW ICE AFFECTS LIFE

Although life persists in glacial ice, one of the most inhospitable environments on Earth, why should you care? After all, most of us don't live in, on, or even in close proximity to ice. Will these organisms affect you, your family, or your neighbor? The quick answer is that they probably can, and they probably already have affected you many times during your life, although you probably did not even comprehend the effects. Research is needed to address the what, when, where, why, and how of life in ice and may provide answers to questions that have perplexed biologists for years; while raising other provocative questions. A few examples follow.

In 1918–1919, the so-called "Spanish" flu (a strain of influenza A virus, which actually originated in China) killed an estimated 50 million people worldwide, many times more than the number of soldiers who died in action in World War I (ongoing at that time); making it the most deadly disease pandemic ever to inflict mankind (at least in recorded history). It killed otherwise healthy young persons, as well as the very young, very old, and the infirm who are the more typical flu victims. Influenza virologists still are not certain why this particular influenza was so deadly although it appeared to have triggered an extreme immune response, causing fluids to build up in the lungs, thus suffocating patients. Some results indicate that genes from two or more different strains recombined to form this deadly strain of influenza. Where did it come from? It disappeared almost as quickly as it appeared, and hasn't been seen since. It is possible that it still exists and might reappear as another pandemic. There is evidence that this can, and indeed has happened with other similar viruses. For example, in 1977, a strain of "Russian" influenza appeared in northern China. It subsequently spread throughout the entire world. This virus strain was almost identical to an influenza strain that caused an epidemic in 1950. Because the influenza virus is highly changeable, it is unlikely to have maintained itself unchanged for 27 years while actively replicating within a susceptible host. One possible, and tacitly accepted explanation by some virologists, is that it was preserved in a frozen state somewhere. But there also is no direct evidence to support this hypothesis. We are currently searching for influenza (and other) viruses in various types of environmental ice. Other pathogens (viruses, as well as bacteria, fungi, and others) also may be in states of suspended animation in glaciers.

When viable viruses and other pathogens of humans, animals, and plants are released into the streams, lakes, and oceans from melting glaciers, they can then move further, being carried by wind, clouds, or animals. This is well documented. In the early twentieth century, it was discovered that bacteria were ejected into the air from water into which they had been seeded. The concentration of bacteria in the aerosol was far higher than that in the water. *Mycobacterium intracellulare*, the cause of Battey infection, can become airborne from the sea. Foot and mouth disease virus, which causes a very serious disease of livestock, is concentrated tenfold by foaming. There is a high correlation between the incidence of viral disease and the use of wastewater for spray irrigation of crops in Israel. Viruses in seawater become adsorbed on air bubbles, and are concentrated and ejected into the air upon bubble bursting. Virus-containing droplets are carried considerable distances by the wind. Several years ago, we (JDC) demonstrated that infectious tomato mosaic tobamovirus (ToMV), a common and widespread plant virus, is present in clouds on the summit of a high peak in the Adirondack Mountains of New York and in fog collected off the coast of Maine. Healthy spruce seedlings exposed to the air but not to native soil in the high peaks of the Adirondack Mountains became infected with ToMV, suggesting that the airborne virus is infectious. The origin of ToMV in clouds and fog is unknown, but because the virus is commonly found in water, it is possible that aerosolized virus from fresh or marine waters is a potential virus source. It is equally likely that the virus melted from the glaciers and snow, since we have found these viruses are well preserved in glacial ice.

Caliciviruses are an interesting group of animal pathogens that present yet another interesting pathway for pathogen movement from sea to land. Many of these viruses probably originated in the oceans. Some of these viruses, most notably vesicular exanthema of swine virus (VESV) and San Miguel sealion virus have wide host ranges that include whales, dolphins, walruses, sealions, fur seals, fish, pigs, donkey, mink, fox, muskox, bison, cattle, sheep, and primates including humans. The marine caliciviruses are capable of producing similar disease in a wide variety of hosts including livestock and humans. VESV is the cause of a severe vesicular disease of pigs that first appeared in 1932 on a swine farm in Orange County, California, where pigs were fed raw garbage, a common practice in those days. The disease spread rapidly so that by the 1950s forty-two states were reporting outbreaks of this new disease. Strains of this virus appear suddenly in animals living in one ocean, and then in subsequent years the same strain will appear in other oceans. While several routes are possible, air, water, and ice may operate in concert to transport this virus temporally and geographically.

In 1992, thousands of people along the coast of Bangladesh began vomiting and experiencing severe dehydrating diarrhea. Many died. The scourge was not new. It was cholera once again, certainly not a surprise to people in this part of the world. What was new was the realization that this epidemic was accompanied by an upwelling of deep-sea water to the surface near the Bangladeshi coast, and the possibility that the cholera pathogen may reside there. Scientists are finding more and more pathogens (e.g., human rotaviruses which cause gastroenteritis), many of which are found only in human feces, surviving at great ocean depths. Do you like to spend a hot day on the beach, with an occasional dip in the lake or ocean? You may want to reconsider. In many shoreline communities of the world, and even in the US, raw or incompletely treated sewage is dumped into these bodies of water. Some ships and boats dump garbage and sewage into the waters of the world. These discharges contain human viral and bacterial pathogens. Infection of even a single person may be enough to start an epidemic of viral or bacterial gastroenteritis or worse. So, there exists a number of mechanisms by which water-borne pathogens of humans, animals, and plants may come onshore to cause disease. Ice may act to preserve these pathogens on a long-term basis, such that they might reappear suddenly after decades (or longer) of absence thousands of miles from where they were originally deposited. Therefore, what is being dumped in the oceans, lakes, streams, and rivers today might cause disease thousands of years from now, thousands of miles away. The next time you travel to the beach or in the mountains, sniff the air as you go. Depending on the weather conditions, you can detect odors from the ocean or lake over long distances, sometimes for hundreds of miles. Then realize that not only are the odors being carried over those distances by the wind, but so are microorganisms. Along with the odors, you inhale hundreds of microbes with each breath. Luckily, most are not human pathogens.

SOURCES AND ADDITIONAL READINGS

Castello, J.D., and S.O. Rogers. 2005. *Life in Ancient Ice*. Princeton, NJ: Princeton University Press.

Castello, J.D., D.K. Lakshman, S.M. Tavantzis, S.O. Rogers, G.D. Bachand, R. Jagels, J. Carlisle, and Y.J. Liu. 1995. Detection of infectious tomato mosaic tobamovirus in fog and clouds. *Phytopathology* 85: 1409-1412.

Dillehay, T.D. 2009. Probing deeper into first America studies. *Proc. Natl. Acad. Sci. USA* 106: 971-978.

Doyle, A.C. 1912. *The Lost World*. London: Hodder & Stoughton.

Gibbons, A. 2014. Three-part ancestry for Europeans. *Science* 345:1106-1107.

Hartl, D.L., and E.W. Jones. 2009. *Genetics, Analysis of Genes and Genomes*, 7th ed. Sudbury, MA: Jones & Bartlett Publishers.

Rassmussen, M., S.L. Anzick, M.R. Waters, P. Skoglund, M. DeGiorgio, T.W. Stafford Jr., S. Rassmussen, et al. 2014. The genome of a Late Pleistocene human from a Clovis burial site in western Montana. *Nature* 506: 225-229.

Reich, D., N. Patterson, M. Kircher, F. Delfin, M. R. Nandinemi, I Pugach, A.M.-S. Ko, et al. 2011. Denisovian admixture and the first modern human dispersals into Southeast Asia and Oceania. *Am. J. Hum. Genet.* 89: 516-528.

Rogers, S.O. 2017. *Integrated Molecular Evolution*, 2nd ed. Boca Raton, FL: Taylor & Francis Group.

Smith, A.W., D.E. Skilling, J.D. Castello, and S.O. Rogers. 2004. Ice as a reservoir for pathogenic animal viruses. *Med. Hypoth.* 63: 560-566.

Templeton, A. 2002. Out of Africa again and again. *Nature* 416: 45-51.

The International HapMap Consortium. 2007. A second generation haplotype map of over 3.1 million SNPs. *Nature* 449: 851-861.

LINKS

https://en.wikipedia.org/wiki/Snowball_Earth
https://en.wikipedia.org/wiki/Spanish_flu
http://www.antarcticglaciers.org/glaciers-and-climate/ice-cores/ice-core-basics/
https://www.researchgate.net/figure/Map-of-Greenland-showing-the-position-of-ice-core-records-red-and-meteorological_fig1_286417720
https://www.sciencedaily.com/terms/antarctic_ice_sheet.htm
https://www.sciencedaily.com/terms/greenland_ice_sheet.htm

2 Questions and Answers

"That is the essence of science: ask an impertinent question, and you are on the way to a pertinent answer"

Jacob Bronowski

ENIGMAS

The enigmas involved in the study of ancient ice are many and varied. The challenges are both large and small, depending on the research question being tackled. They will certainly keep people studying, thinking, and hypothesizing for a very long time. Many basic questions still are completely or partially unanswered. Imagine if you could go back in time and view the world from a perspective that only ancient humans and hominins could have had. Go back 200,000 years, to a time when our species (*Homo sapiens*) was a new species. Go back even further to a time that predates *Homo sapiens*. What did the world look like then? What sorts of organisms were there? What kinds of diseases plagued our human and hominin ancestors, other animals, and plants at that time? What sorts of geological processes were occurring at that time? Were there any large bolide impacts on the Earth? In more recent times, what was occurring? What was occurring during the times of human population increases, decreases, migrations, and industrialization? How did microbes affect human populations and their migrations?

In the past four decades, researchers (including ourselves) have begun to understand some of these processes through the study of ice and microorganisms in ice. This includes ice in glaciers, permafrost, and in ice fields and ice domes. When microorganisms, including bacteria, archaea, fungi, viruses, and other microbes become entrapped in ice, they may enter a state of suspended animation, or they may actually metabolize and reproduce in ice. They also are entrapped with soil, meteorites, volcanic dust, plants, animals, gases, and other particles, thus saving a snapshot of the atmosphere at various points in time. Through close examination, the events occurring on the Earth hundreds to millions of years ago can be reconstructed.

FROZEN KNOWLEDGE

There are several interesting and important scientific implications to survival of organisms in ice. Global climate change has been the subject of countless scientific papers and newspaper stories for more than three decades. Global warming causes increased glacial melting, which results in accelerated releases of viable microbes from them. This may increase the chance of disease outbreaks in populations of humans or other organisms (animals, plants, microbes, etc.). By analogy, the effects of pollution may be causing some detrimental effects that were not predicted even a few decades ago. Thus, some organisms may be good indicators of global (or local) change, and some may benefit other organisms. If such sentinel organisms can be found and identified, they may be useful to scan the ice for past events as well as to indicate current local or global problems. The key here is to find organisms that indicate problems before they have become untreatable or irreversible, as well as to find the beneficial organisms before they are gone.

A related implication can be illustrated using smallpox virus as an example. This virus was eradicated from human populations in 1979. In other words, this virus was forced into extinction (although some samples are being retained for research purposes in a few repositories around the world). Attempts are being made to eliminate other viruses, including poliovirus, from the Earth. However, these attempts might be impossible for those that can survive freezing. For example, smallpox virus and poliovirus both remain viable after freezing and thawing. If these organisms survive entrapment and release from glaciers, then it is possible that we will see them return. Viruses with sturdy protein coats are the most likely to survive for long periods of time in glaciers. However, some viruses that are enveloped by host cell membranes also can survive freezing, as can many bacteria, archaea, and eukaryotes. Influenza A viruses are surrounded by host membranes, but they survive freeze-thaw cycles, with little loss of viability. The only way to determine if they pose any threats is to study these microbes in detail, including their susceptibility characteristics during freezing and thawing, and their prevalence in glaciers.

Fossils are remnants of organisms from past geological ages that are embedded in rock or other materials for very long periods of time. Certain organisms are sometimes called "living fossils" (e.g., crocodiles, monitor lizards, coelacanths, dawn redwoods, ginkgo, and others). However, this is not accurate, because each of them has never been a fossil. The crocodiles in Africa and the ginkgo trees growing in this country are not dead or dormant. They metabolize, reproduce, and respire. They are living now. The word "fossil" has been broadened to mean a species that has been on Earth for a very long time. However, the microbes that survive being frozen for long periods in ice can be accurately called "living fossils." They have halted growth and reproduction, sometimes for millennia, but once thawed, they grow and reproduce under appropriate conditions. This may be one of the reasons that microbes are one of the most diverse sets of organisms on Earth. Not only have they had a longer time to diversify, many are more resistant to episodes of mass extinction events than are the large multicellular organisms (such as ammonites, dinosaurs, trilobites, humans, and others). This suggests another distinct advantage of being able to survive in ice. If a large ecological disaster (e.g., volcanic eruptions, meteor impact, nuclear war, etc.) occurs while the organisms are frozen in the ice, they may be able to avoid extinction. Therefore, during each mass extinction event that has occurred on Earth, a large number of microbial species may have survived, or even thrived, by hiding out in ice (or in deep water), thus maintaining microbial diversity.

EVOLUTIONARY IMPLICATIONS

Viruses are essentially mobile infectious pieces of DNA or RNA encased in a protective coat. They are capable of reproduction if they infect the appropriate host and can also introduce some or all of their DNA (or a DNA copy of their RNA) into the host chromosomal DNA. This means that an ancient virus that has recently emerged from a glacier can carry ancient genes and insert them into a modern organism. This is not a crazy idea. Horizontal transfer of genes (transfer of genes from one organism to another, whether or not they are the same species), mediated in part by viruses, is rampant, and has been going on throughout the entire history of life on Earth. In fact, it has been responsible for many major evolutionary changes. Depending on the genes inserted, this can affect growth rates, survival, evolutionary rates, and other characteristics. Some viruses carry genes that can initiate cancers in the infected organisms. At this time no one knows whether these viruses are capable of surviving entrapment in glaciers, but given the diversity of viruses on Earth, it is likely that some do. Therefore, from evolutionary and medical respects, it is essential to examine the potential preservation of these microbes in ancient ice.

One reason we began research on organisms in ice is that we thought it would be possible to measure evolutionary rates of change by comparing DNA and RNA sequences among individuals at various points of time. That is, we thought that the organisms and their genes were frozen in time and we could therefore reconstruct the molecular evolutionary changes through time. However, we soon discovered that ancient and modern organisms and their genomes (the total complement of genes per cell) are constantly mixing. Thus, a gene that might have been prevalent in the population thousands of years ago may again become established in a modern population through the process of entrapment in and release from glaciers. This confuses measurements of evolutionary change, since there is always a mixture of organisms that have undergone different degrees of mutation based on the amounts of time that they have been metabolically active and reproducing. This process, which we call, "Genome Recycling" is discussed in detail in a later chapter.

The Cambrian Explosion (which occurred about 540 million years ago) was a time of rapid divergence of animals. There was more animal diversity at that time than exists today! No one knows its exact cause. One theory is that it may have occurred as a result of a worldwide ice age, known as the "snowball Earth hypothesis," which occurred from approximately 800 to 600 million years ago. Thick ice extended from pole to pole, including equatorial regions. During this time, organisms may have been separated by large ice fields, essentially creating islands of life throughout the world. Islands are hotspots for speciation because groups of organisms are physically separated by large distances. Thus, during this extensive ice age, speciation events may have increased. When the ice melted, all of the new species could expand their ranges and diverge further as they migrated to new environments. In addition, when the ice melted, it undoubtedly released enormous quantities of microbes with the potential to speed rates of speciation and evolution, not only of the microbes, but also rates for other organisms, due to horizontal gene transfer events, as discussed above. A related hypothesis to the "snowball Earth hypothesis" is that life on Earth had its origins in cold environments rather than hot ones (the predominant hypothesis). There is compelling evidence to support both hypotheses, including combinations of both.

APPLICATIONS

In addition to the interesting biological and evolutionary aspects of this research, there are also practical benefits to the study of microbes in ice. Viable isolates of ancient bacteria, archaea, or fungi might be capable of producing anticancer drugs, antibiotics, or enzymes and other biomolecules that can be used for biomedical or industrial uses. Most of the biomedical products and food additives currently in use originate either from microorganisms or from plants. The biomedical field and related industries worldwide spend billions annually on the development and refinement of useful biological products, including new enzymes used for a broad range of purposes, the genetic identification of people and cancers, diagnosis of infectious diseases, and indicating when toxins are present. Yet standard enzymes for these applications have physical limitations. There are many research groups in the world actively searching for microorganisms that live in extreme environments (extremophiles) and their enzymes (extremozymes) for industrial applications. It was a thermostable DNA polymerase (*Taq* DNA polymerase) isolated from the thermophilic bacterium (*Thermus aquaticus*) that made the revolutionary polymerase chain reaction (PCR) technology feasible. PCR is a method whereby one or a few specifically targeted DNA molecules from environmental (or other) sources can be used to produce millions of copies in only a few hours. It is used by thousands of scientists and technicians every day. We use this method to detect microorganisms in the extreme environment of ancient ice.

Similarly, another heat-loving extremozyme in commercial use increases the efficiency of cyclodextrin production from cornstarch. Cyclodextrins are used to stabilize volatile flavorings in food, improve uptake of medications by the body, and to reduce bitterness of food and medicine. Cold-loving organisms (psychrophiles) also interest industrial research scientists, who need enzymes that function at low temperatures for use in food processing at refrigerator temperatures, fragrance makers whose products volatilize at higher temperatures, and producers of cold-wash laundry detergents. Other currently untapped useful biological products undoubtedly exist in these ancient microbial populations.

EXTRATERRESTRIAL LIFE

While no one can yet say that the study of microbes in ice will resolve the question of the origin of life on Earth, there are some intriguing possibilities. Water ice has now been found on many bodies in our Solar System (e.g., the Moon, Mars, Ceres, Mercury, Pluto, and some of the moons of Jupiter, Saturn, Uranus, Neptune, and Pluto), and may exist beyond our tiny corner of the Universe. If water is a universal requirement for life, as it is here on Earth, then life may have had multiple sites of origin. If ice is as good a preservative on other bodies as it is on Earth, then microbes and organic compounds are likely to be found in these places, as well as traveling through space encased in ice. If microorganisms or their nucleic acids are found on Mars or Europa, there are at least two possibilities: 1. Life originated on Mars or Europa, or 2. Life began on Earth and traveled to the other bodies by being blown out of Earth's gravity by a volcanic event or a meteor collision. Thus, life may have originated somewhere else, and only became deposited on Earth about 4.2 billion years ago.

Perhaps life has been deposited on the Earth several times in the past, and we may be continually bombarded by extraterrestrial life. Tons of water and dust particles are being added to the Earth daily from comet and other space debris. Microorganisms may be included in this debris. Thus, Earth may be constantly inoculated with extraterrestrial organisms. There is a growing interest in this area of research, being one aspect of the field of Astrobiology. If these organisms originally came from Earth, of course, they would simply be returning home after long journeys. However, if they originated elsewhere and have been seeded on Earth, then Earth's biota may be an amalgam of organisms that originated in many different locales in the Solar System and beyond. Life may have existed on Mars or Europa millions or billions of years ago. If so, ancient life may be preserved in deep sediments or ice on these extraterrestrial bodies or on Earth. The technology and protocols developed to detect life in deep ice cores could be used on other worlds to detect life there. Most of this technology is in its infancy, with much more to come. With any luck, a space probe may soon detect unequivocal signs of life on another planet or moon in our Solar System. While this might solve some mysteries, it will likely generate others.

MOVING FORWARD

The bacterial cell that just entered your lung may have had a very long journey. It may have hitched a ride on a piece of soil. If it was embedded in a rock encased in an icy comet or simply on a piece of space dust, who knows where in the Universe it came from? Is this how our Earth was originally populated by living organisms? Or did life originate on Earth and migrate to other planets and bodies in our Solar System and beyond? Many viruses participate in horizontal gene transfer events (i.e., transfer of genes from one species to another). Could they accomplish these transfers over very long distances (i.e., between planets), or on long time scales? In this book, we will try to answer these questions, although in many cases we only scratch the surface and will pose more questions than answers. We hope, however, that the relevance of discovering life in ancient ice will become clear to you. It is our hope that you will begin to ask new questions of your own. In Chapter 1, we asked whether it was possible to revive the dead. We know now that the answer is that it is possible to revive the apparently dead, because they were not yet dead, just in a resting state. The important questions to address that need answers are: 1. How long can organisms be preserved in ice?; 2. How can living organisms (or signs of life) be detected in ice?; 3. How can organisms be revived from ice?; 4. Are some dangerous?; 5. Are some beneficial?; and 6. What will happen once the ice is gone?

SOURCES AND ADDITIONAL READINGS

Augustin, L., C. Barbante, P.R. Barnes, et al. (55 authors). 2004. Eight glacial cycles from an Antarctic ice core. *Nature* 429: 623-628.

Castello, J.D., and S.O. Rogers. 2005. *Life in Ancient Ice*. Princeton, NJ: Princeton University Press.

Cheng, J., J. Abraham, Z. Hausfather, and K.E. Trenberth. 2019. How fast are the oceans warming? *Science* 363: 128-129.

Petit, J.-R, J. Jouzel, D. Raynaud, N.I. Barkov, J.-M. Barnola, I. Basile, M. Bender, J. Chappellaz, M. Davis, G. Delaygue, M. Delmotte, V.M. Kotlyakov, M. Legrand, V.Y. Lipenkov, C. Lorius, L. Pépin, C. Ritz, E. Saltzman, and M. Stievnard. 1999. Climate and atmospheric history of the past 420,000 years from the Vostok ice core, Antarctica. *Nature* 399: 429-436.

Rogers, S.O. 2017. *Integrated Molecular Evolution*, 2nd ed. Boca Raton, FL: Taylor & Francis Group.

LINKS

https://climate.nasa.gov/
https://en.wikipedia.org/wiki/Global_warming
https://en.wikipedia.org/wiki/Smallpox
https://www.space.com/35469-solar-system-habitable-icy-worlds-infographic.html

3 The Importance of Water and Ice

"Without ice, the earth will fall"

Emma Thompson

LIQUID AND SOLID WATER

Water is the solvent required for all life, at least life as we know it. All organisms contain a great deal of water, usually around 70%, some more, some less. It interacts with all of the molecules in cells, and is used for transport of chemicals, and is involved in many of the biochemical reactions in living organisms. Water is a fairly simple compound, containing two hydrogen atoms and one oxygen atom. However, it has some very special properties. It has a positively charged end (where the hydrogens are) and a negatively charged end (where the oxygen is). This causes the molecules to line up in very specific ways, and when frozen, the semicrystalline structure of molecules causes solid water to be less dense than liquid water. This is why ice floats atop liquid water. It also means that when icebergs calf from the glaciers, they float off into the surrounding oceans and lakes, sometimes traveling for very great distances. Ice is simply frozen water, but the structure and attraction of the molecules makes it a unique matrix for survival of microorganisms and their biomolecules, and depending on the temperature and pressure during freezing, water molecules can pack into at least 18 different geometries, as well as amorphous forms, which causes it to have a diversity of characteristics.

Often, when I (SOR) am discussing issues of molecular biology and microbiology with students, I will say, "Water isn't water, and ice isn't ice." After some confused looks, I explain. What I mean is that a glass of tap water is very different from the ultrapure water that we use in our research laboratory, and sea water is very different than lake water. The ultrapure water has almost no ions and no organic carbon (indicating the amount of contamination by molecules that originated from organisms), and we go further by autoclaving the ultrapure water (heating to 121°C under pressure) to assure that no living things or active biological molecules exist in the water. Typical tapwater contains dissolved ions, particulate matter (e.g., sand, plastic, mud), microorganisms (including bacteria, fungi, protists, viruses, and others), and many chemicals. There are large numbers of microbes in your tap water. Even if it is treated in your area, by the time it reaches your tap, it is loaded with microbes, and even after it is frozen large numbers of viable microbes remain (Fig. 3.1). This type of water cannot be used for any molecular or microbial research, because some chemicals inhibit the enzymes and other constituents that are used in molecular biology methods, and the concentrations of microbes are often much higher in tapwater than in the environmental ice samples. Even bottled distilled water is not pure enough for this research.

To show you how vital ultrapure water is for this research, a few years ago, the molecular reactions that usually worked every time in my research and teaching labs were failing. We could not figure out what was causing these failures. We tried ordering chemicals from different companies, recalculating all of the formulas for reagents, and I asked different students to try different parameters, but the reactions failed more often than not. Then, I passed a graduate student from another lab in the hallway one summer day and asked her how things were going. She said that she was very disappointed, because the bacterial cultures that had always grown well in the lab, were now only growing slowly, or not at all. So, I asked other labs about their experiments. Most were having major problems with simple things that had previously always worked. The final clue was one day when I asked a new undergraduate student to make a solution of EDTA (ethylene-diaminetetraacetic acid, a chemical that is commonly used in molecular biology labs). But, I did not tell him that the pH had to be above 8.0 for the chemical to dissolve in the water. Usually, sodium hydroxide is used to increase the pH so that the EDTA will go into solution. When I arrived in the lab with a smile on my face,

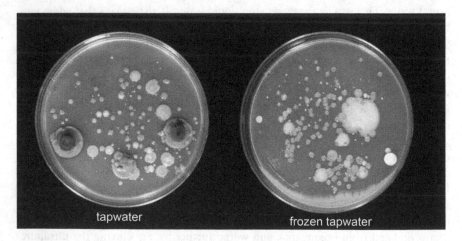

FIGURE 3.1 Fungi grown from 200 µl (0.2 ml) of tap water (left). More than 100 colonies can be seen, presumably each originating from a single viable cell, representing at least a dozen species. The culture plate on the right was inoculated from 200 µl of the same water after it had been frozen at -20°C and then thawed. The number of colonies is similar, but the colony types and diversity differ.

I asked whether he had made the solution, expecting him to say that the EDTA would not dissolve. Instead, he said, "Yes, it dissolved with no problem." I was shocked and confused. So, I then tested the pH of the pure water supply (reverse osmosis purified water) coming from taps throughout the building, and found that instead of being slightly acidic (pH 5–6, which is normal for this water), it was alkaline, with a pH of 9.5! Then, I spread some of the water from various locations in the building onto agar plates with bacterial and fungal growth media. A myriad of bacteria and fungi grew. The "pure water" supply in our building was anything but pure. After multiple and expensive attempts to clean the system, new pipes were run through the building, with pumps that recirculated the water constantly through filters to assure purity. The take home lesson is that all water is not the same, and even so-called pure water is not always very pure.

WHAT IS IN THAT WATER?

Years ago, one of us (SOR) was flying back to the US from South Africa (after a stint in Swaziland, now Eswatini, with the US Peace Corps), by way of Rio de Janeiro, Brazil. The flight from Cape Town to Rio was uneventful. After stopping in Rio for several hours, we boarded another plane for the flight to Miami. The plane was absolutely full. Carnival in Rio was underway. From the first part of the flight I noticed something strange. More than the normal number of people were lined up at the toilets, a few were groaning, and the others were sleeping, somewhat restlessly. What was going on? The couple next to me explained what had happened. It turned out that nearly the entire plane was full of shoe salesmen and saleswomen from the Miami area who had decided to go to Rio for Carnival. While in Rio, they had a great time for the first day. They all were very careful not to drink the water, settling for bottled water and beer most of the time. However, at a party on the second night they all had mixed drinks, and all of the drinks contained ice. The ice apparently was contaminated with some microorganism that gave all of them an enteric disease. By the next day, the merry-making had ceased, and most of them spent two days or more with severe diarrhea. By the time they boarded the plane, more than half were still ill. The shoe salesmen and saleswomen had assumed what many assume, that freezing kills disease-causing organisms. That assumption is wrong. And this has been known for more than a century. Your tap water, whether you drink it right out of the tap, or freeze it and then drink it, contains a multitude of viable microbes (Fig. 3.1). Ice is an excellent preservative of microbes, nucleic acids, and other biological constituents. As an example, in our labs we constantly store biochemicals (including enzymes) and organisms (including bacteria, viruses, fungi, and others) in -20°C or -80°C freezers.

WHY ARE WATER AND ICE SO IMPORTANT FOR LIFE?

Most biological molecules (specifically, DNA, RNA, and proteins) must have shells of water around them to function. Water molecules are part of the structure of biological molecules. The interactions between these molecules and water are vital to their structures and functions. Additionally, many of these molecules require co-factors, usually small molecules or ions (charged versions of atoms and molecules). These small molecules must be able to diffuse through the water in order to attach to the large biological molecules. The biological molecules include those that build other biological molecules (synthesizing enzymes), and those that break them down (degradative enzymes), as well as nucleic acids. The degradative enzymes must be removed or kept inactive in order to preserve these biological molecules. For example, in methods for DNA purification, the enzymes that degrade the DNA (known as nucleases) are first inactivated by removing their co-factors (generally magnesium ions), and then the enzymes are removed by using an organic solvent (such as phenol or chloroform) and subsequent precipitation of the DNA out of the water-soluble fraction. In natural tissue preservation, inactivation of the degradative molecules occurs through freezing and/or desiccation. When frozen, the water molecules form rigid shells that immobilize the DNA, proteins (including enzymes), and co-factors for the enzymes. When tissues are desiccated, the water surrounding the biological molecules and co-factors is absent, such that the rate at which they move and interact with one another is greatly reduced or completely stopped, depending on how much water is removed. When you buy dried foods (meats, fruits, etc.) they contain much less water than the fresh versions, and can remain stable for very long periods of time. The DNA in those foods can be preserved for moderate amounts of time (probably for months to years, or longer). In mummifications, there is less water, but there is some that still is present, because the shells that are closest to the biological molecules are the most difficult to strip off.

Natural preservations are generally of better quality than those performed by humans (Fig. 3.2). Seeds preserved in packrat middens often are more than 50,000 years old. Packrat middens are mainly composed of fossilized packrat urine with embedded plant materials. Packrats bring seeds and leaves back to their den and urinate on them. The urea eventually crystallizes and builds up to form a urea matrix surrounding the plant materials. Urea attracts water to itself (i.e., it is hygroscopic), and thus acts to desiccate the seeds and leaves, naturally mummifying the plant materials. [Maybe humans should take a lesson from the packrats and start urinating on their samples.] Specimens in amber (fossilized plant resins) are sometimes more than 100 million years old. Amber is impervious to almost everything, including water and oxygen. Salt also is a natural preservative. There are reports of the isolation of living bacteria from 250 million-year-old salt deposits. As with urea, salt draws the moisture away from the cells and thus dehydrates them.

Humans have engaged in tissue preservation, and have done so for thousands of years. The first ones that come to the minds of most people are the mummies of Egypt. Various salts and chemicals were used, as well as removal of the internal organs, including the brains, which were homogenized and then sucked out through the nose. Although there is excellent preservation of tissues near the surface of these corpses, the deeper tissues often are severely degraded. Additionally,

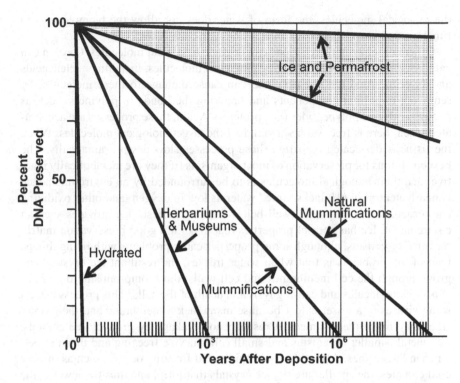

FIGURE 3.2   Rates of DNA degradation under various conditions. When hydrated, DNA is rapidly degraded by enzymes, oxidation, and other chemical processes. When tissues are dried or desiccated, as in mummifications and herbarium preparations, the DNA can remain relatively intact for hundreds to thousands of years. Natural mummifications are more effective in DNA preservation than human attempts at mummification. Ice and permafrost combine cold temperatures and dehydration (because of the lack of free water), which will preserve the DNA for tens of thousands to millions of years.

there are herbaria and museums worldwide where humans have preserved plants, fungi, animals, and bacterial cultures. When kept cool and dry, these organisms and their biological molecules can be preserved for centuries. Humans also have dried and frozen foods for millennia. Both of these methods work well, primarily because the degradative processes, active when the tissue is hydrated, are greatly slowed down due to the lack of water. At first, it seems a paradox to think of ice as a matrix that dehydrates cells and tissues, but frozen water is very different than freely moving liquid water. As we said above, all biological molecules need to be surrounded by water to function. However, this includes not only the water molecules with which they are in direct contact, but also includes so-called "free" water of the inner shells surrounding the biological molecules. Thus, in ice there is very little "free" water available, because the water is locked as a rigid matrix, and the degradative molecules cannot reach their target biological molecules. Therefore, even though there are many water molecules surrounding the biological molecules, the form of water does not allow the biomolecules to function, and keeps them from interacting with each other.

Other non-enzymatic reactions can degrade biological molecules. Oxygen can cause oxidative breakdown of many biological molecules, including nucleic acids and proteins. Acid (or low pH) also can cause damage to these molecules by removing important components and breaking the bonds in polymers, such as nucleic acids and polypeptides (i.e., proteins). Again, these processes are accentuated when there is free water surrounding the target biological molecules. When the tissue is desiccated or frozen these processes slow down dramatically. The best conditions for preservation of most organisms (if they are metabolically inactive) and their biological molecules is to be surrounded by an environment that is nearly neutral in pH, lacks "free" water, is low in oxygen (and other oxidizing chemicals), and is very cold (well below freezing is best). Ice provides such an environment. Ice has special properties that make it a good preservation matrix for microorganisms, although some properties cause problems with living things. One of the problems is that when water freezes, the resultant ice crystals can grow through the cell membranes and cell walls, thus compromising the gradients of biomolecules and causing lysis and death of the cells. This process affects some viruses to a lesser extent because many lack a membrane and their small sizes may make them somewhat resistant to breakage by growing ice crystals. In general, smaller organisms and small cells survive freezing and thawing better than larger ones. Also, the more rapidly that freezing occurs, such as in very cold climates, the smaller are the ice crystals that form, and thus the lower is the potential for cell damage.

ICE AS A MICROBIAL ARCHIVE

Ice in glaciers and ice sheets has been preserving organisms and biological molecules for hundreds of millions (perhaps for billions) of years. Organisms, large and small, have been caught in snow and rainstorms, blown in sea foam, fallen into crevasses, and multitudes have become entrapped in glaciers. This has included all sorts of plants, animals (including humans), and microbes, and has been occurring since ice has existed on the Earth and probably elsewhere in the Solar System and beyond. Thus, deposition of ice has become a sampling mechanism for various periods in Earth's history. The layers in a glacier represent a variety of components from the atmosphere that were present for varying amounts of time. Each depth, then, represents an average of conditions and microbes over a period of time depending on the rate of snowfall, and the time when the ice becomes compacted enough to be isolated from the atmosphere. Some organisms die when they are frozen. For example, no one has yet brought back to life a frozen human or mammoth, although they have extracted intact DNA from both. On the other hand, many organisms, both unicellular and multicellular, survive multiple freeze-thaw cycles. For example, organisms as large as mosses, algae, insects, amphibians, crustaceans, and others can survive being frozen.

The number, types, and diversity of microbes in the ice are dependent on many factors. Snow and ice usually require a nucleating factor to begin the crystallization process. Often, this is a single microbe, many microbes, pollen, or a dust particle. The dust particles also may carry microbes. The number of microbes in the snow that falls, and the dust and sea foam that settles on the glaciers and other surfaces depends on the geographical location, the weather, and what microbial sources are upwind of the ice surfaces. For example, ice near coastal areas in the Arctic and Antarctic receive snowfall from clouds that contain microbes from the oceans, and from larger organisms, as well as from sea foam that blows across the land. Ice from these areas usually contains high concentrations of microbes, often exceeding tens of thousands of microbes per milliliter of meltwater, and high proportions of the microbes remain viable. Far inland, it is a different story. In areas near high human habitation, such as the Alps and the Himalayas, moderate to high numbers of microbes are found in the ice and glaciers. On the other end of the spectrum, in the very remote areas of Greenland and Antarctica, the concentrations of microbes in the ice can be as low as a few cells per milliliter of meltwater, and in the deepest ice, not only are the concentrations of microbes low, but very few of them are still viable. Some microbes may prove important in allowing us to predict what might happen in times of climate change. Others may prove important in causing human disease, or in aiding in human health. Only additional studies of microbes in ice will provide us with the information to determine the impacts of the preservation of microbes in ice, as well as the consequences of the increases in the rates of melting worldwide.

PRESERVATION OF MICROBES IN GLACIERS

Glaciers have different cycling times (i.e., the time it takes for something that is entrapped in the glacier to be released at the terminus, or from the surfaces) because of the diversities in glacial flow rates and lengths. Therefore, some microbes are only entrapped for a few years, while others may be entrapped for hundreds of thousands of years, and in some cases, for more than a million years. Some organisms may exit the glacier by surface melting of the glacier. They can be picked up by the wind or ingested by another organism, or they may be refrozen in the same glacier, or in other glaciers. Microbes entrapped in ice generally lose viability over time (Fig. 3.3). However, many have ways to repair their cells, and especially their DNA. Some bacteria are metabolically active at temperatures down to at least -80°C. This means that some organisms are likely able to continue basic housekeeping functions to maintain cell viability. However, this ability apparently is not unlimited, because studies of microbial viability in ice cores has demonstrated that the numbers of viable cells decrease continuously with depth (i.e., age) in the glaciers.

Ice appears to be an ideal medium for preserving organisms in an historical context. It has several features not shared by most other ancient substrates. Glacial ice is a natural air-sampling matrix. Any airborne organism or particle has the potential to fall onto ice fields in snow or dust, and eventually become

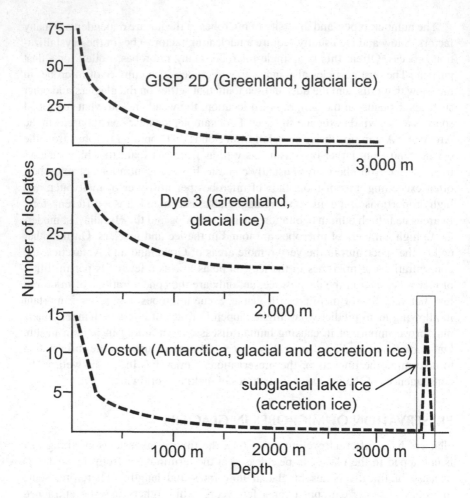

FIGURE 3.3 Graphs showing the rapid loss of viability measured in a few ice cores from Greenland and Antarctica. In general, as the ice hardens and is compacted, the entrapped microbes tend to die, although some of the hardy and adaptable species can survive even at extreme depths and pressures. In the case of accretion ice from Lake Vostok, a rise in the number of viable cells indicates the existence of organisms living in the lake water.

part of a glacier. Thus, glacial ice provides a snapshot of what was present in ancient atmospheres, whether it is dust particles, microorganisms, pollen grains, atmospheric pollutants, cosmic dust, or other inclusions. There are tremendous numbers of viable organisms in the ice and in the meltwater from the ice. Based on our culture results, there are from a few to hundreds of viable organisms per milliliter of Arctic and Antarctic meltwater. Based on melting rates measured over the past two centuries, we estimate that the release of viable microbes world-wide is approximately 10^{18} to 10^{22} cells annually. Recent increases in the rate of melting worldwide due to the effects of greenhouse gases are increasing the rates of release of viable microbes by orders of magnitude. In temperate and tropical glaciers, there may be higher numbers of viable microbes melting from the ice. Also, in glaciers with short cycling times (from freezing at the top to thawing at the terminus), there are more viable organisms released by melting at the end of the glacier or from other surfaces. Thus, there can be as many as hundreds of thousands of viable microorganisms per liter of glacial ice meltwater. A large portion of the human population relies on glacial meltwater for drinking, as well as for agriculture. Few know that there are large numbers of viable microbes in this water, many of which are plant and/or animal pathogens.

MICROBES AND ALIENS

Microbes constantly are floating in the atmosphere. You inhale dozens of bacteria, fungi, viruses, and other microbes with every breath. Many are stuck to dust particles. A few may exist as free particles. Dust particles are the nucleation sites for ice crystals in hail and snow. Bacteria, fungi, protists, and viruses (as well as larger organisms, or parts thereof) adhere to dust and other particles, which become parts of snowflakes and raindrops. These fall into oceans, streams, and lakes, as well as onto land and glaciers (and onto your hair, skin, clothing, and pets). If conditions are conducive, these snowflakes and raindrops become frozen into glaciers. Because microorganisms entrapped in ice are released when the ice melts, the organisms contained within are recycled into the biosphere, a phenomenon unlikely to occur with ancient subterranean life forms. Thus, organisms that are able to survive freezing and thawing in ice can lengthen their life spans and recycle their genomes through geological time scales. About 3% of the Earth's surface (10% of the land area) is covered by ice year round. An unbelievably huge number of microbes (and other larger organisms) are entrapped in this ice, with almost 10^{18} to 10^{22} viable microbes (possibly more) melting out of the ice annually. Even if a very small proportion of these are human pathogens, their releases represent potentially significant world health issues. And human pathogens are only a small percentage of the variety of pathogenic microbes that exist on Earth. Other animals, plants, and virtually all organisms are susceptible to a variety of microbial pathogens. Therefore, release of microbes from glaciers is an important issue from many standpoints.

The process by which these organisms become airborne and then deposited also may have caused some to escape the Earth and to travel through space. Whether they survive space travel is still a matter of speculation. However, space is a very cold place, as long as direct sunlight is avoided (e.g., by shading). Microbes have been found floating kilometers up in the atmosphere. Viruses and bacteria have survived experimental exposure to space. The length of time they may survive is unknown, although there are anecdotal reports of survival at least for years. Nevertheless, some (whether living or dead) undoubtedly have landed on other celestial bodies. Some of these bodies have much more ice than the Earth. The polar environment on Earth is believed to be a reasonable surrogate for present conditions on Mars, several moons of Jupiter and Saturn, Pluto, comets, and other locations in the Solar System and beyond. Even if life does not exist on these bodies today, it may have existed in the past and is perhaps preserved in ancient ice there. That life may have originated on Earth and been blasted off of the Earth by asteroids and/or volcanoes, followed by space travel on ice and dust, to finally land on another body within the Solar System or beyond. Alternatively, life may have originated elsewhere, and by the same process it seeded the Earth. Two such theories have been proposed on this subject. Panspermia states that life is spread throughout the Universe, and Earth was seeded by some of these organisms. Exogenesis is a bit more conservative, stating that life originated somewhere in the Universe, and was transferred to Earth, but that it may or may not be widespread in the Universe (possibly only in a single locale). These two theories are receiving more attention by biologists working with ancient ice cores. If viable life is preserved for long periods under such conditions on Earth, by analogy it is reasoned that the same might be true on these extraterrestrial bodies. A comparison of ancient life forms on Earth with those potentially present on other planets may someday lead to new theories concerning the origin of life on Earth and in the Solar System. But ice is the key, and where the focus should be for seeking ancient life on Earth and elsewhere.

SOURCES AND ADDITIONAL READINGS

D'Elia, T., R. Veerapaneni, and S.O. Rogers. 2008. Isolation of microbes from Lake Vostok accretion ice. *Appl, Environ. Microbiol.* 74: 4962-4965.

D'Elia, T, R. Veerapaneni, V. Theraisnathan, and S.O. Rogers. 2009. Isolation of fungi from Lake Vostok accretion ice. *Mycologia* 101: 751-763.

Ma, L., C. Catranis, W.T. Starmer, and S.O. Rogers. 1999. Revival and characterization of fungi from ancient polar ice. *Mycologist* 13:70-73.

Ma, L., S.O. Rogers, C. Catranis, and W.T. Starmer. 2000. Detection and characterization of ancient fungi entrapped in glacial ice. *Mycologia* 92: 286-295.

Rogers, S.O. 2017. *Integrated Molecular Evolution*, 2nd ed. Boca Raton, FL: Taylor & Francis Group.

Rogers, S. O., L.-J. Ma, Y. Zhao, V. Theraisnathan, S.-G. Shin, G. Zhang, C.M. Catranis, W.T. Starmer, and J.D. Castello. 2005. Recommendations for elimination of contaminants and authentication of isolates in ancient ice cores. In: Castello, J.D., and S.O. Rogers (eds.) *Life in Ancient Ice*. Princeton NJ: Princeton University Press, pp. 5-21.

Rogers, S.O., Y.M. Shtarkman, Z.A. Koçer, R. Edgar, R. Veerapaneni, and T. D'Elia. 2013. Ecology of subglacial Lake Vostok (Antarctica), based on metagenomic/meta-transcriptomic analyses of accretion ice. *Biology* 2: 629-650.

Rogers, S.O., V. Theraisnathan, L.-J. Ma, Y. Zhao, G. Zhang, S.-G. Shin, J.D. Castello, and W.T Starmer. 2004. Comparisons of protocols to decontaminate environmental ice samples for biological and molecular examinations. *Appl. Environ. Microbiol.* 70: 2540-44.

Shtarkman Y.M., Z.A. Koçer, R. Edgar R, R.S. Veerapaneni, T. D'Elia, P.F. Morris, and S.O. Rogers. 2013. Subglacial Lake Vostok (Antarctica) accretion ice contains a diverse set of sequences from aquatic, marine and sediment-inhabiting Bacteria and Eukarya. PLoS ONE 8(7): e67221. doi:10.1371/journal.pone.0067221.

LINKS

https://en.wikipedia.org/wiki/Glacier
https://en.wikipedia.org/wiki/Ice

4 The Diversity of Ice

"Ice contains no future, just the past, sealed away"

Haruki Murakami

SNOW AND ICE DIVERSITY

Native peoples who live in Northern Canada and Siberia have dozens of names for snow and ice, depending on its characteristics. Environmental ice varies dramatically from one location to another, from one season to another. If you live in regions with snow, or have spent some time skiing, traveling through snowy areas, or hiking on glaciers, you already have seen the evidence. Upstate New York and other downwind areas around the Great Lakes, are prone to bouts of heavy lake effect snow. Clouds form over the lakes during the fall and winter when cold air flows over the relatively warm water of the lakes. The snow-laden clouds then drop the snow downwind over the land. Once the snow is two feet or more in depth, you can feel the weight when you shovel your driveway or rake the snow off of your roof. If you wait a day or two, the snow is more compacted and dense, and often has an icy crust on top due to surface melting. If there already was snow on the ground before the heavy snow, it has usually been partially melted and refrozen, and if a car has run over it, or a snowplow has tossed it onto your driveway, then it is a jumble of ice chunks, snow, and slush that is difficult to dig through.

The roads leading to ski areas often have high snow banks surrounding the sides of the roads. Layers of snow, representing the sequential snowfalls can be seen, and compaction of the snow can be observed, similar to what happens in glaciers, but on a small scale. Most ski areas broadcast the skiing conditions for the day. This is partly a weather forecast, but also they detail what type of snow the skiers can expect. In the Fall, the snow is usually wetter and heavier, but the ski areas "groom" most of the ski runs, by driving over them with rollers that compact and smooth the snow, so that skiing is easier. By mid-winter, there is more powder snow, which is lighter, fluffier, and is what many advanced skiers love. Spring skiing is dicey. Some of the ski runs more exposed to the sun can be mainly slushy snow, which has a high water content. Runs that are partly shaded can be slushy in the sun, but icy in the shade. So, part of the time you would be trudging slowly through slushy snow in the sun, and then suddenly hit a very slippery patch of glare ice in the shade. Often, these are areas where skiers are most likely to fall.

Some of the same conditions exist in environmental ice, including, glaciers, ice fields, sea ice, ice domes, and icebergs (Fig. 4.1). When the snow falls early

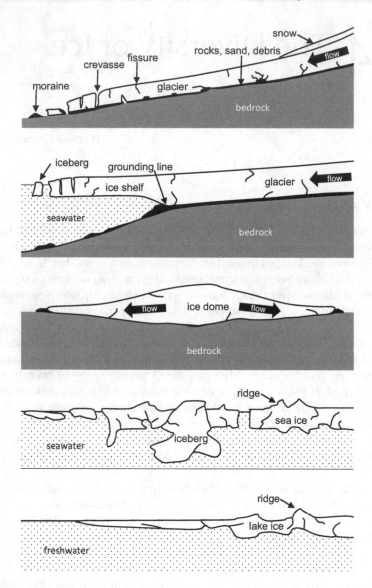

FIGURE 4.1   Structure of ice systems (from top to bottom: glacier, glacier/ice shelf, ice dome, sea ice, lake ice). In glacier systems, snowfall is compacted into ice, which creates the glacial ice, some of which interacts with the underlying bedrock, producing basal ice, which contains embedded rocks, sand, and other inclusions. Ice shelves are extensions of glaciers as they flow into bays and oceans. They are composed of freshwater ice, although seawater also can accrete to the lower surfaces of the ice shelf. Ice domes occur in areas where the net flow is relatively slow, although they have outlet glaciers that continuously release ice from the domes. Sea ice can form in patches that eventually grow and adhere to other patches. Some of these interactions form ridges between the two patches that are forced together. Some freshwater glaciers become entrapped as the sea ice forms around them. Lake ice usually begins at the edges of lakes and continues to the middle. Ridges and cracks also form in similar ways to sea ice, especially in large lakes.

in the winter, it is usually wet and heavy, especially at low elevations, but later is light and fluffy. In the spring, when it has partially melted, it is slushy, heavy, and dirty. By the middle of spring, the mounds of snow look more like piles of debris. Heavy wet snowflakes form in wetter and warmer clouds, whereas small snowflakes form in drier and colder clouds. This is because the crystals form slowly in warm, wet clouds, so that they have longer times to grow. In drier, colder clouds, the crystals form rapidly, and most of the moisture is depleted from the areas surrounding the newly formed crystals, so they remain small. Much of East Antarctica is a cold desert, where the snow consists of small crystals, and annual snowfall accumulation rates are low. The fringes of Antarctica near the coastlines, West Antarctica, and the Antarctic Peninsula receive more snowfall, with much greater annual accumulation rates. The crystal sizes within glaciers can indicate weather conditions that were present when the snow was deposited.

GLACIERS

As alluded to above, all ice is not the same. Snow differs from glacial ice, and the top layers of glacial ice differ from those in the middle and on the bottom of the glacier. Even snow comes in different forms. Ice that forms on surface lakes differs from ice that forms over subglacial lakes. Ice can take different forms at different pressures and temperatures. At temperatures just below the freezing point, water crystals can grow very large, especially when under pressure. However, at very low temperatures, the crystals are very small. Also, as water freezes, it does not form an orderly crystal, as do other chemicals. While some of the water molecules line up in a regular fashion, others are more randomly arranged, which is one of the factors that causes snowflakes to vary from one to another, and makes solid water less dense than liquid water, which causes ice to float on top of liquid water. While it is generally assumed that water freezes into ice at 0°C, this is not always true. Water can be supercooled to well below 0°C without freezing. However, if the supercooled water is seeded with a small crystal of ice, or a bit of dust, or it is vibrated slightly, it will freeze throughout immediately.

Each year of snowfall forms distinct layers, due to freezing, melting, and packing. Thus, ice cores can be accurately dated. Deeper in the glaciers, the pressures compress the ice, and layers cannot be observed directly. However, isotopes of gases trapped in the ice can be analyzed to deduce dates for specific depths. Presently, some cores have been dated to over one million years using these methods. Therefore, a chronological sequence can be generated of what was present in the atmosphere over discrete periods of time. Also, it may be possible to correlate the presence of specific microorganisms, (e.g., pathogens or thermophilic microorganisms) with specific historical (e.g., plagues), geological (e.g., volcanic eruptions), or climatic (e.g., periods of global warming or cooling) events in Earth history. The chronological sequence preserved in ice cores is one method by which climatologists can accurately determine climate change over time. Glacial ice may be unique in this respect.

Ice at the bottom of a glacier, called basal ice, differs significantly from ice higher in the glacier (Fig. 4.1). This is because the ice interacts with the subtending bedrock. You can think of it as being similar to a river or stream moving over a rocky surface. The ice is cracked, broken, pulverized, tumbled, and mixed with particles of the rock that are broken off by the forces from the glacial movements. This interrupts the layers in the glacier, such that the age of the basal ice cannot be precisely determined. Also, the microbes are almost all nonviable, and when they have been examined microscopically, they show signs of being crushed and sheared into pieces. While viability of organisms generally decreases with depth in the main portions of glaciers, it appears to drop even faster in basal ice.

Glaciers and other types of ice appear in all zones of the world, covering roughly 10% of the land surface (or about 1.5 million square kilometers, Table 4.1). The ice within or near the Arctic (primarily Greenland) and Antarctic (Antarctica and surrounding islands) comprises about 98% of the freshwater ice on Earth. But glaciers also exist at lower latitudes all the way to the equator. Being in such diverse areas, you might expect that there are different types of glaciers, and there are. There are fast-moving glaciers, slow-moving glaciers, glaciers that end in the sea, and those that end as a stream, river, or spring. Some form on the edges of cliffs and are thus generally very short, while others are thousands of miles in length. About 84% of all terrestrial ice on Earth is in Antarctica. Another 14% is in the Arctic, most of which is in Greenland. The remaining 2% occurs on the mountains of the world, with the majority in the Himalayas. There is only a tiny bit near the equator, primarily in the Andes in South America, and on Mt. Kilimanjaro and Mount Kirinyaga (formerly Mt. Kenya), both in Africa. And, because of global climate change, that ice is melting rapidly. As you read this, the permanent ice in Africa might already be gone.

Glaciers range from being relatively smooth rivers of ice to fields of ice blocks. They may be receding or advancing, although currently over 75% are receding.

TABLE 4.1
Geographic Distribution of Ice on Earth

Region	Area Covered by Ice (km²)	Percent of Total Land Area Covered	Total Ice Volume (km³)	Percent of Total Ice Volume
Antarctica	12,588,000	84.1	29,500,000	91.4
Greenland	1,803,000	12.1	2,600,000	8.1
Other Arctic	335,000	2.2	93,600	0.29
Asia	117,000	0.8	34,000	0.11
North America	77,000	0.5	21,300	0.066
South America	26,500	0.2	8500	0.027
Europe	7400	0.05	2100	0.0066
South Pacific	1200	0.008	320	0.0010
Africa	12	0.00008	3	0.00010
TOTAL	15,000,000	100	32,260,000	100

In the Alps, accurate records have documented the recent mass recession of the glaciers. They may meet with other glaciers or be more or less isolated. They generally have gravel fields, or moraines, on their flanks and at their termini. There are short and long glaciers, as well as those that are so short that they are little more than ice bowls formed on the side of a mountain. Some glaciers may have streams at their centers. Others have water streams that flow on top or through them and many have surface pools during the warmer months. They do not all flow and retain the layered pattern present at their source. The ice may form waves and even tumble over like an ocean wave. Again, the river model can be used. If you follow a water molecule in a river, it may start out on the bottom and then be thrust to the top, and then down again. In a glacier similar movement may occur, but to a far smaller extent, and on a smaller scale. In solid ice, the water molecules do not move vertically for the most part. The microbes are carried by the ice, but may be able to move in microscopic liquid water veins as well. Where ice crystals meet other ice crystals, there may be minute veins of liquid water, due to high concentrations of solutes. It has been hypothesized that this is a specialized habitat where some microbes may be able to live, metabolize, and reproduce. However, this remains to be demonstrated.

Ice in glaciers differs somewhat from ice in ice fields and domes (Fig. 4.1). Sea ice and lake ice are very different from both of these. The most obvious difference is that snow and ice are deposited on the upper surfaces of glaciers, ice fields, and domes, while sea ice and lake ice (including accretion ice overlaying subglacial lakes) freezes to the lower surfaces of the existing ice. In glaciers, snow falls in the upper regions of the glacier, and the weight of more snow causes compaction of the underlying snow (Fig. 4.2), and also causes the glacier to move downhill. Early in this process, the layers are still in contact with the atmosphere, so there are exchanges of gases, and if there is rain, snow, glacial surface streams, or glacial surface lakes, some of this can percolate into the lower layers. This is called the firn layers, or simply, firn. Eventually, approximately 50–100 m lower, the snow is compacted into dense ice that no longer is in contact with the atmosphere, and the amount of air in the ice is diminished. However, it contains trapped atmospheric gases, and these can be used to determine the age of the ice at various depths, the temperature of the atmosphere at the time of deposition, and the concentrations of CO_2, methane, and other gases. The bottom layers of glaciers cease to be in layers, because the ice interacts with the bedrock, and tumbles, folds, grinds, and cracks, in a similar way that water on the bottom of a stream or river interacts with the stream and river bed. Thus, the age of the ice, gas concentrations, etc., are difficult to determine in these regions. Also, the viability of cells in these regions is lower than in the overlying intact layers, and the cells appear very distorted when observed microscopically.

The ice, and its gas inclusions are under pressure. As ice core sections are brought to the surface, and are warmed at all, the trapped gases can expand, and the ice can fracture and/or explode. In most cases, when shallow to mid-depth ice core sections are brought to the surface, the ice core sections are stored below freezing for weeks or months, before they are examined and studied. During the

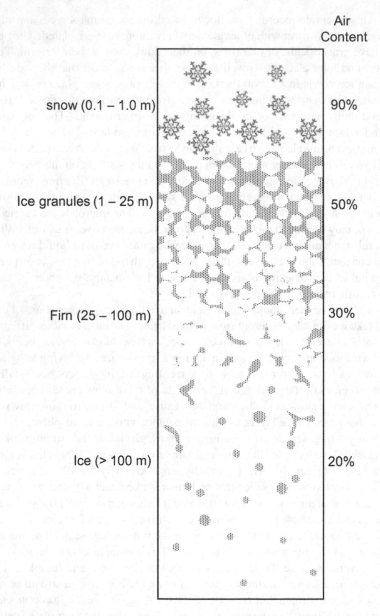

snow (0.1 – 1.0 m)

Ice granules (1 – 25 m)

Firn (25 – 100 m)

Ice (> 100 m)

Air Content

90%

50%

30%

20%

FIGURE 4.2 Conversion of snow into ice within a glacier or ice dome. Snow is approximately 90% air. As more snow falls, the underlying snow becomes compacted into ice granules, where the snow crystals have been compacted and space between each of the granules leads to less air space. Further compaction results in the firn layers, with the reduction of airspaces. While the firn may extend to 50–100 m below the surface, some connections with the atmosphere remain. Finally, approximately 100 m below the surface, the ice and airspaces are compacted to the point where there is no longer any connection between the ice and the atmosphere.

resting period, some of the water molecules reorient, making the ice core stronger, such that it can withstand the pressures of the expanding gases. Deeper ice experiences even higher pressures, which causes further compression and changes in the ice structure, producing ice that can withstand the pressures of the gases. Microbes also are squeezed. In fact, it is amazing that so many survive, given the high pressures, cold, low levels of nutrients, lack of free water, and darkness.

Glaciers also are unequal in their rates of ice accumulation. Accumulation rates at Byrd station, in West Antarctica, are high, because it is under marine influences and the glaciers in the area accumulate ice rapidly. At the other extreme, Vostok, Antarctica lies at high altitude in East Antarctica, within a frozen desert, and receives very small amounts of annual snowfall due to its position on the leeward side of the Transantarctic Mountains. Accumulation of ice by glaciers here is exceedingly slow. The concentrations of microbes are also very low. Glaciers incorporate different sets of microorganisms or mix them in different ways through the years because there are so many types of glaciers. Therefore, the blend of microbes melting from the flanks of Mt. Kilimanjaro is likely to be very different than that melting from a glacier in East Antarctica. What this means is that worldwide the variety of sites for the study of microbes encased in ice is virtually unlimited.

As mentioned above, glaciers are diverse. Some are short, while others are kilometers long. For example, the Gem Glacier in Glacier National Park covers only 5 acres of land, while the Lambert Glacier in Antarctica is 435 × 96 km in length and width (at the widest point). Some are only meters thick, while some are several kilometers thick. For example, the glacier that flows over Lake Vostok (Antarctica) is more than 4 km thick. Glaciers often are cracked and uneven. Most have dust, sand, or stones on top, as well as deeper within, and at their bases. Some glaciers have a significant number of rocks, pebbles, sand, and boulders. Glaciers deposit rock, sand, and debris, along their margins, called moraines. Lateral moraines are formed on the sides of glaciers, while terminal moraines are formed at their ends. Medial moraines form when two glaciers converge, and subglacial moraines form on the bottoms of the glaciers where there is often a stream flowing beneath the glacier. The deposits and streams contain rock, sand, solutes, debris, remnants of dead organisms, and many viable microbes. These vary depending on the glacier, and the position within the glacier. Many moraines that formed during the last ice age can still be seen as hills and ridges throughout the world. Many of these probably contain some of the organisms that were deposited by those glaciers.

Some glaciers contain ice that is millions of years old, while others contain ice that is only a century or two old. Some carry huge amounts of silt and organisms entrapped in the ice. These are usually in mountainous regions that also have frequent rain, snow, dust, and wind storms. Glaciers flow at different rates. Some move slower than 200 meters per year, while some flow at rates of more than 40 meters daily! Recently, some Greenland and Antarctic glaciers have begun to move faster. Time-lapse videos of some of the fastest glaciers closely resemble liquid rivers. There are regions that look like rapids, eddies, and pools. Thus, although they appear static, they are dynamic systems. They move, carry soil, rocks, and microbes, and of course they deliver a constant source of water to the land and life below. The glaciers in the Himalayas supply water to about one-fourth of the human population (approximately 2 billion people). One wonders whether any of these people realize that they are drinking water that contains millions of ancient microbes.

ICE FIELDS AND DOMES

Ice fields and ice domes move differently than glaciers. They often flow outwards (downslope), and feed into one or more glaciers (Fig. 4.1). Conceptually, one can think of ice systems as slow-moving water systems, with lakes, streams, and rivers. Ice fields and domes are analogous to ponds and lakes. However, many of these systems, also have liquid lakes, streams, and rivers on their surfaces during the summers and at their lower surfaces (subglacial lakes, streams, and rivers). During summer in Greenland, Antarctica, and elsewhere, large lakes form on the surfaces of the glaciers, and some disappear within hours, as they break through cracks and crevasses in the glacier and flow down huge waterfalls through the glacier, eventually reaching the base of the glaciers. This volume of water lubricates the base of the glacier, which sometimes causes an increase in the flow rate of the glacier. There is usually input of snow and ice from surrounding areas, as well as from precipitation, causing rises in the levels of the ice fields and domes. But, there also are regions of outflow, and for ice fields and ice domes, these are glaciers. One difference between lakes and ice domes is that the dome can indeed be much higher in the center than around the edges. This causes high pressures under the ice, and in general, there is a flow of ice outwards, and towards the outlet glaciers. One advantage for scientists studying inclusions in ice (e.g., gases, ions, microbes) is that the ice can occur in very regular layers, without a great deal of distortion, which is more characteristic of glaciers because they interact with bedrock, and twist and turn on their movement towards their termini. Several ice cores have been obtained from ice domes, both in Greenland and Antarctica, which have yielded very precise results.

LAKE AND SEA ICE

Sea ice differs from freshwater ice. The dissolved salts cause differences in the way the ice crystals form in sea ice. It freezes at about -3°C, and can be sliced through by freshwater glacial ice, which is harder. While the ice in glaciers, ice fields, and ice domes builds up from bottom to top, ice in the oceans and lakes freezes from the top to the bottom (Fig. 4.1). Therefore, the age profiles are reversed. For example, the Vostok (Antarctica) ice core consists of several types of ice. The top 3,300 m is glacial ice, which contains layers of ice ranging in age from the present to 400,000 ybp. From 3,300 m to 3,538 m, the ice is disorganized and jumbled. This is the basal ice. It has been estimated that the ice at 3,538 m is from 500,000 to 1 million years old. At 3,539 m and below, the ice originated as water from the large subglacial Lake Vostok. This ice is up to 10,000 to 15,000 years old (the time it takes for each section of the glacier to move over the width of the lake). The ice is called "accretion ice." The ice at the surface of the water is recently accreted ice. Therefore, the age profile for the glacial ice, starting from the top, is from most recent to most ancient, while the accretion ice is from most ancient to most recent.

Lake and sea ice can remain for only a season or in some cases for decades. As stated above, the ice freezes from the top to the bottom, so the most active areas are usually at the interface between the ice and the water. Sea and lake ice have more diverse ecosystems than other types of ice. The lower portions of the ice consist of crevices, cracks, and caverns that are habitats for a diversity of bacteria, algae, microscopic animals, and protists. Enough sunlight often penetrates through the ice such that photosynthetic organisms are active. All of this biological activity leads to the formation and release of nutrients that are utilized by planktonic organisms that become the food for larger animals. On the upper surfaces of the ice, there also are a variety of microbes, including cyanobacteria and other photosynthetic organisms. In addition, the upper portions of the ice entrap microbes transported by wind and precipitation. Together, there can be thousands of different species either entrapped or living on and in lake and sea ice.

Sea ice comprises an extremely large and diverse ecosystem. It covers approximately 7% of the surface of the Earth, or 12% of the ocean surface. It is composed of a mix of solid salt water ice, fractured ice, pancake ice, ice ridges, and icebergs (consisting of freshwater ice calved from glaciers and ice shelves). The thickness varies, and there are cracks, pockets of water, channels, and other features that present a diverse set of habitats for microbes (including bacteria and archaea), animals (including fish and crustaceans), fungi, protists (including diatoms and algae), plants, and other organisms. Additionally, there are sets of organisms that live on the water side of the ice, while there are other sets that live on the upper surfaces of the ice. The ice can vary from less than one year old to more than centuries old, or older (in the case of embedded icebergs). The salinity differs from newly formed ice that is close to the salinity of sea water to ice that is denser, and has forced out some of the salt, to the glacial ice that is mainly freshwater. It can contain viable organisms, as well as fragments of dead organisms of varying ages.

SURFACE AND SUBGLACIAL LAKE ICE

Lakes in polar regions can remain ice covered for decades, while others may freeze and thaw annually. We investigated surface ice from eight Antarctic lakes and found that there was a great deal of microbial diversity, primarily dependent on both distance from the ocean and proximity to human habitation. The lake that had the greatest number and diversity of microbes and the greatest diversity was near Bellinghausen station (established in 1968), which is close to the ocean, on an island off the Antarctic Peninsula. A large number of these microbes may have been deposited in the lake ice by blowing sea foam. Ice from the other lakes contained fewer microbes and lower species diversity. The lowest was from a lake far from the ocean and any settlements. Siberian lakes and many lakes in temperate regions freeze annually. Influenza A viruses were identified in two of three Siberian Lakes that we assayed for the virus. These were breeding grounds for migratory waterfowl. In the case of Lake Erie ice, a metagenomic study indicated that many of the microbes were those expected in the lake, including large numbers of cyanobacteria and diatoms, as well as other microbes characteristic of lake environments. However, a large proportion of the sequences were from organisms that indicate human influences, including sequences from agricultural species, cooling water from industrial plants, and human pathogens.

Subglacial accretion ice is different in many ways from other types of ice. Where the overriding glacier is partially grounded on the lake bottom, silt and sand are entrapped in the accretion ice, in addition to any microbes in the lake water. Over the lake water, the ice is extremely clear, with almost no silt or sand inclusions. Both can also include entrapped gases and ions. Lake Vostok accretion ice contains a high concentration of ions (i.e., salts) in the location of a shallow bay that might have nearby hydrothermal activity. The sodium, chlorine, and other ions are in concentrations similar to brackish water in estuaries, where freshwater mixes with sea water. Further out in the main lake basin, the ion concentrations are close to those of pure water. Microbe concentrations fell to almost zero in these regions. Also, because the temperature is close to 0°C, ice crystals can grow to very large dimensions. Some ice monocrystals have been measured at more than one meter in length! When these crystals contact adjacent large ice crystals, they form what are called triple junctions, because three crystals meet along their edges. As they grow, the crystals push solutes along their fronts, and at the triple junctions, the solutes are concentrated. Liquid water can be present in these regions, which might provide nutrients and suitable environments for certain microbes. Lake Whillans is a small shallow subglacial lake (under 800 m of ice) in West Antarctica that was the subject of recent study. Researchers drilled into the lake and took samples from the lake water and sediment cores from the bottom of the lake. The ecosystem consisted mainly of methanogenic archaea that had been present for approximately 120,000 years. While there were some similarities between the microbial community represented in Lake Vostok and Lake Whillans, there also were major differences. The details are discussed in a later chapter.

SOURCES AND ADDITIONAL READINGS

Augustin, L., C. Barbante, P.R. Barnes, et al. (55 authors). 2004. Eight glacial cycles from an Antarctic ice core. *Nature* 429: 623-628.

Castello, J.D., and S.O. Rogers. 2005. *Life in Ancient Ice*. Princeton, NJ: Princeton University Press.

Christner, B.C., J.C. Priscu, A.M. Achberger, C. Barbante, S.P. Carter, K. Christianson, A.B. Michaud, J.A. Mikucki, A.C. Mitchell, M.L. Skidmore, T.J. Vick-Majors, and the WISSARD Science Team. 2014. A microbial ecosystem beneath the West Antarctic ice sheet. *Nature* 512: 310-313.

Petit, J.-R, J. Jouzel, D. Raynaud, N.I. Barkov, J.-M. Barnola, I. Basile, M. Bender, J. Chappellaz, M. Davis, G. Delaygue, M. Delmotte, V.M. Kotlyakov, M. Legrand, V.Y. Lipenkov, C. Lorius, L. Pépin, C. Ritz, E. Saltzman, and M. Stievnard. 1999. Climate and atmospheric history of the past 420,000 years from the Vostok ice core, Antarctica. *Nature* 399: 429-436.

Rogers, S.O., Y.M. Shtarkman, Z.A. Koçer, R. Edgar, R. Veerapaneni, and T. D'Elia. 2013. Ecology of subglacial Lake Vostok (Antarctica), based on metagenomic/meta-transcriptomic analyses of accretion ice. *Biology* 2: 629-650.

Shtarkman Y.M., Z.A. Koçer, R. Edgar R, R.S. Veerapaneni, T. D'Elia, P.F. Morris, and S.O. Rogers. 2013. Subglacial Lake Vostok (Antarctica) accretion ice contains a diverse set of sequences from aquatic, marine and sediment-inhabiting Bacteria and Eukarya. PLoS ONE 8(7): e67221. doi:10.1371/journal.pone.0067221.

Zhang, G., D. Shoham, D. Gilichinsky, S. Davydov, J.D. Castello, and S.O. Rogers. 2006. Evidence for influenza A virus RNA in Siberian lake ice. *J. Virol.,* 80: 12229-12235.

LINKS

https://en.wikipedia.org/wiki/Glacial_motion
https://en.wikipedia.org/wiki/Ice_sheet
https://en.wikipedia.org/wiki/Retreat_of_glaciers_sincc_1850
https://en.wikipedia.org/wiki/Sea_ice
https://nsidc.org/cryosphere/glaciers/questions/located.html
http://www.antarcticglaciers.org/glaciers-and-climate/ice-cores/ice-core-basics/
http://www.antarcticglaciers.org/glaciers-and-climate/shrinking-ice-shelves/ice-shelves/
https://www.livescience.com/47461-lake-whillans-species-antarctica-life.html

5 Permafrost

"The universe is like a safe to which there is a combination, but the combination is locked up in the safe"

Peter de Vries

FROZEN SOILS

Permafrost consists of soils, sands, clays, and/or rocks that are frozen for at least two consecutive years. Approximately 24% of the land area not covered with permanent ice, in the Northern Hemisphere consists of permafrost (Table 5.1). More than 64% of this is in the Tibetan Plateau and the Atlai Range that straddles Mongolia and Russia. Almost 22% is in Siberia, Alaska, and Greenland (again, the areas not covered by permanent ice). Some permafrost has remained frozen for centuries or millennia, or longer, while some experiences occasional thawing. For many regions, the upper layers experience temperature variations and partial thaws, but lower layers remain frozen at a constant temperature below freezing. Just as soils are diverse, permafrost is diverse. Some permafrost is fairly dry, although the water that is present is frozen. Some is primarily fine sand, while others are made of thick clay. Some permafrost is mostly frozen bogs. Many contain pockets of water, some of which are the size of small puddles, and others that are the size of ponds and lakes. Permafrost also exists on sea beds under ice shelves.

LIFE IN PERMAFROST

There are many different types of permafrost that form a variety of habitats for microbes (and other organisms), from mostly terrestrial to mostly aquatic, and even marine habitats, as well. As expected, microbial diversity in permafrost is high, but there are some organisms found in permafrost that were not expected. While it was once thought that everything in the permafrost was dead, we now know that a number of microbes, as well as some multicellular organisms, survive for very long periods of time within permafrost. A diverse set of organisms occur in permafrost. Viable bacteria that function in the nitrogen cycle have been isolated from permafrost as cold as -50°C. Carbon-fixing bacteria also have been recovered from permafrost. In essence, communities of bacteria and other microbes exist in permafrost that form complete ecosystems capable of complex metabolic activities. A diversity of fungi also has been isolated from permafrost.

TABLE 5.1
Extent of Permafrost by Geographical Region

Location	Area (km²)	Percent of World Total
Tibetan Plateau	1,300,000	36.4
Atlai (Mongolia/Russia)	1,000,000	28.0
Brooks Range (Alaska)	263	7.4
Siberian Mountains	255	7.1
Greenland	251	7.0
Ural Mountains	125	3.5
Andes Mountains	100	2.8
Rocky Mountains	100	2.8
Fennoscandia Mountains	75	2.1
Other	<100	2.8
TOTAL	3,569,000	

Interestingly, the species of bacteria and fungi are similar to those that have been isolated from ice cores from the Arctic and Antarctic. The same species that have become adapted to cold environments may be able to colonize several different cold habitats. Some of the permafrost that has been studied is up to 8 million years old, suggesting that these ecosystems have very long histories.

In the early 1990's, two groups of scientists began examining the lung tissues of corpses from graves that were dug in permafrost decades earlier. These people died of influenza A during the 1918 pandemic. The scientists were trying to ascertain why the 1918 influenza A strain, from the H1N1 subgroup, was so virulent. By studying the RNA sequences from the influenza A chromosomes, it was hoped that the reason for the virulence of this strain could be deduced. It was important to find lung tissues that showed signs of influenza A infection, and tissues that had been well preserved during the decades of burial. The two groups began exhuming bodies from the graveyards in two small towns thousands of miles apart: Spitsbergen, Norway and Brevig Mission, Alaska. The two towns were similar in that both were built on permafrost. This meant that the corpses had remained frozen in ice for decades. As it turned out, the corpses in Spitsbergen had not always remained frozen, since some of the permafrost had thawed from time to time over the years. But, the corpses in Brevig Mission had remained frozen, and the 1918 H1N1 influenza was successfully isolated and sequenced (although with great difficulty). While the sequences yielded some clues about the virus, the two studies did demonstrate that microbes are preserved in permafrost. Attempts were made to revive the viruses. Ultimately, or happily, none were viable. However, the sequences themselves told a story. The 1918 influenza A pandemic was caused by a strain of H1N1 that had pieces of its genome from humans, pigs, and birds; but was mainly a human strain. Some had hypothesized that it was a bird flu strain that had somehow become adapted to infect humans. Had that been the case, it might have been even more dangerous and deadly, because no human would have had any immunity (either complete or partial) to that strain.

In 2016, in a remote region in Siberia, there was an outbreak of a devastating disease that killed more than 2,000 reindeer, and dozens of people were hospitalized. One child died. The pathogen was confirmed as *Bacillus anthracis*, causative agent of anthrax. Where had it come from? This area was far away from any other settlement, and travel into and out of the area was very limited. However, this region had experienced a serious outbreak of anthrax 75 years before, and recently, the permafrost in the region had been warming and thawing more than in previous decades. Spores of *B. anthracis* are known to be resistant to many types of degradation, and they survive freezing and thawing. A close relative of this bacterium is *B. subtilis*, and its spores can survive for thousands of years. Therefore, it was concluded that the recent permafrost thawing released some viable bacteria that had infected animals from the previous outbreak, which infected the reindeer, which then infected the humans, producing the outbreak. Fortunately, the disease was identified, the people were treated and evacuated, and the outbreak was stopped before it spread further. Will the disease emerge again as more permafrost melts? Probably.

In 2017, researchers isolated and grew viruses recovered from Siberian permafrost that was 30,000 years old. These viruses (*Pithovirus sibericum* and *Mollivirus sibericum*) only infect amoeba, so humans do not have to worry, but other viruses may be preserved in the permafrost. In the 1890s, there was a smallpox epidemic in Siberia. The bodies of the dead were buried in permafrost near the banks of the Kolyma River, in northern Siberia. Recently, the Kolyma River has been slowly eroding its banks near the graves, and it is unknown how hazardous the situation will become if some of the corpses are exposed by the melting permafrost. Smallpox virus does survive freezing and thawing in laboratory and environmental settings, and freezing preserves many types of viruses quite well. So, although it was declared in 1979 that smallpox had been eradicated from the Earth, some scientists are now questioning whether this is 100% certain. While the viruses can survive freezing and thawing, it is unknown how long the viruses survive after being frozen for so many years in these corpses. That the amoeba viruses survived for 30,000 years should at least make us vigilant to the possibility of viral diseases as the permafrost melts and exposes potential pathogens.

In 2018, the revival of worms (small nematodes) from old Siberian permafrost was reported. More than 300 of these worms were found, of which few were alive. But, two of them, one from 32,000-year-old permafrost, and one from 42,000-year-old permafrost, not only moved, but began eating. They appeared to be fully functional multicellular organisms. Approximately one-third of their cells are neurons, responsible for responding to the odors of food, moving muscles, making decisions, and sensing environmental stimuli. It is quite amazing that these animals not only could crawl, but they could find and move toward food and eat it! So, not only can viruses, bacteria, fungi, and other microbes survive being frozen for millennia (and longer), but there are several reports of the revival of complex multicellular organisms from ancient frozen matrices. The microscopic organisms known as water bears (i.e., tardigrades) survive for very long periods of time in extremely low temperatures (down to almost absolute zero), and once thawed, they revive, eat, and reproduce as if nothing had happened.

In Alaska, Russia, and Greenland, many villages and cities are in danger because of thawing permafrost, and coastal and river erosion. The permafrost is shifting, cracking, and slumping, such that some houses, buildings, and roads are sinking, tilting, and being destroyed. Temperatures in the Arctic (and Antarctic) are increasing at double and triple the rates of those in temperate and tropical regions. The rising temperatures cause the permafrost to thaw and slump, and then in the winter, it refreezes and heaves upward. The houses, roads, and buildings are subjected to constant movement with the seasons, which makes building and maintaining them difficult. The population may be exposed to the living pathogens thawing from the permafrost, which might potentially lead to major disease outbreaks. Currently, the risk of disease in these regions is unknown, primarily because it has not been investigated in a focused way, extensive thawing of the permafrost is a relatively recent phenomenon, and the populations in the areas are small. The transmission of pathogens released from permafrost will be studied in greater detail.

Toxins as well as pathogens are locked in the permafrost. Recently, it was discovered that large amounts of mercury are frozen in the permafrost. The Arctic permafrost alone may contain more than 15 million gallons of mercury. If there is large scale thawing of the permafrost, the release of large amounts of mercury into the environment could have devastating consequences for animals, plants, and humans. Much of it could enter the oceans, which could cause widespread illness and death in the microbes, plants, and animals in the oceans. Portions of it would also be concentrated in fish that would be eaten by other animals, and by humans. In addition to acute toxic effects of mercury, long-term exposure to mercury causes kidney failure and neurological disorders. The amount of mercury discovered in permafrost is larger than the amounts found in any other land or ocean environment.

Living cyanobacteria have been detected in permafrost. Cyanobacteria are photosynthetic bacteria that are found almost everywhere on Earth, but are primarily known for their ecological and hazardous characteristics in aquatic and marine systems. Some of the permafrost in which they have been detected has been buried for millions of years. This is surprising because light does not penetrate more than a centimeter or so into the permafrost and cyanobacteria are photosynthetic. How do they survive in the permafrost? It turns out that many cyanobacteria can also grow as heterotrophs, being independent of light, but instead obtaining their organic carbon compounds from other sources. Viable cyanobacteria have also been found deep in glacial ice, subglacial lakes, and deep water where the amount of light is too low for photosynthesis. As in permafrost, these organisms are able to grow and divide as heterotrophs in the complete absence of light. It is important to realize that many organisms are very adaptable, including cyanobacteria. Even *E. coli*, commonly thought to be only a strictly human gut bacterium, can live in a wide variety of environments and live on many different types of nutrients. This also is a characteristic of organisms that live or survive encased in ice and permafrost.

Many large mammals went extinct approximately 10,000 to 11,000 years ago, coinciding with human migration and increased predation on these animals. However, some remnants of these mammals are preserved in permafrost, including specimens of Woolly mammoth (*Mammuthus primigenius*). This species went extinct on continental regions approximately 10,000 years ago. However, two populations, one on St. Paul Island, Alaska, and one on Wrangel Island, Russia, survived until 5,600 and 4,000 years ago, respectively. Some of the tissues were in good condition, such that DNA sequences could be determined for both the mitochondrial and parts of the nuclear genomes. Based on these sequences, this species was most closely related to Asian elephants. Efforts are underway to resurrect the species, using the DNA sequences already obtained, and a surrogate Asian elephant for gestation. However, significant biological and ethical concerns remain.

RELEASE OF GREENHOUSE GASES AND OTHER CHEMICALS

As permafrost melts, many microbes within it begin to metabolize more rapidly. Some microbes begin to eat dead organisms or other organic matter. Both processes release large amounts of methane and carbon dioxide. Also, the thawing permafrost releases trapped gases, including more carbon dioxide and methane, as well as water vapor, all of which are greenhouse gases. Water vapor becomes a greenhouse gas when it forms haze and clouds. Methane is a potent greenhouse gas, 84 times more potent than carbon dioxide, although it is around for a shorter time in the atmosphere. However, methane reacts with hydroxyl groups to produce carbon dioxide and water, again contributing to the overall content of greenhouse gases in the atmosphere. Furthermore, the amount of carbon dioxide emitted into the atmosphere is about eight times that of methane.

Carbon dioxide is being added to the atmosphere by burning fossil fuels, which causes heating of the atmosphere, land, and water. The heat is generated when sunlight strikes land and water, and is converted into heat, which radiates into the atmosphere. However, the carbon dioxide blocks the heat from escaping the atmosphere, and it is reflected back to the land and water. When the sunlight penetrates water, especially oceans, again it is slowed and dissipated as heat, which warms the water. Much of the sunlight energy is being absorbed by the oceans, and the excess heat, which is trapped by atmospheric carbon dioxide (and other greenhouse gases) also is heating the oceans, much more so than the land. This causes more evaporation of water from the oceans, which exacerbates the problem by further insulating the Earth. Water molecules have an average residence time in the atmosphere of 9 days, methane has a residence time of 7 years, and carbon dioxide has a residence time of 35 to 90 years.

There is a positive feedback loop to all of this. The more the permafrost melts, the more carbon dioxide, water vapor, and methane are released, which causes more heat to be trapped by the atmosphere, which leads to more heating and more melting. The increased warming of the oceans and atmosphere also lead to more melting of glaciers. Several estimates of the loss of glaciers in the Himalayas suggests that from 35% to 67% of the ice in the Himalayas could be gone by 2100. Approximately one-quarter of the 7 billion people on Earth depend on these glaciers for their drinking water, agriculture, and industry. If this water were to disappear from the Himalayas, agriculture would fail, economies would crash, diseases would spike, and many would die. The area could not sustain the current number of residents if half of this water supply disappeared. This is discussed in more detail in a later chapter.

SOURCES AND ADDITIONAL READINGS

Faizutdinova, R.N., and D.A. Gilichinsky. 2005. Yeasts isolated from ancient permafrost. In: Castello, J.D., and S.O. Rogers (eds.) *Life in Ancient Ice*. Princeton NJ: Princeton University Press, pp. 118-126.

Ivanushkina, N.E., G.A. Kochkina, and S.M., Ozerskaya. 2005. Fungi in ancient permafrost sediments of the Arctic and Antarctic regions. In: Castello, J.D., and S.O. Rogers (eds.) *Life in Ancient Ice*. Princeton NJ: Princeton University Press, pp. 127-139.

Legendre, M., A. Lartigue, L. Bertaux, S. Jeudy, J. Bartoli, M. Lescot, J.-M. Alempic, C. Ramus, C. Bruley, K. Labadie, L. Lhmakova, E. Rivkina, Y. Couté, C. Abergel, and J.-M. Claverie. 2015. In-depth study of Millivirus sibericum, a new 30,000-y-old giant virus infecting Acanthamoeba. *Proc. Natl. Acad. Sci. USA* 112:E5327-E5335.

Revich, B.A., and M.A. Podolnaya. 2011. Thawing of permafrost may disturb historic cattle burial grounds in East Siberia. Glob. Health Action 4: 10.3402/gha.v4i0.8482.

Rivkina, E., J. Laurinavichyus, and D.A. Gilichinsky. 2005. Microbial life below the freezing point within permafrost. In: Castello, J.D., and S.O. Rogers (eds.) *Life in Ancient Ice*. Princeton NJ: Princeton University Press, pp. 106-117.

Taubenberger, J. K., A.H. Reid, R.M. Lourens, R. Wang, G. Jin, and T.G. Fanning. 2005. Characterization of the 1918 influenza virus polymerase genes. *Nature* 437:889-893.

Vishnivetskaya, T.A., L.G. Erokhina, E. V. Spirina, A.V. Shatilovich, E.A. Vorobyova, A.I. Tsapin, and D.A. Gilichinsky. 2005. Viable phototrophs: cyanobacteria and green algae from the permafrost darkness. In: Castello, J.D., and S. O. Rogers (eds.) *Life in Ancient Ice*. Princeton NJ: Princeton University Press, pp. 140-158.

LINKS

https://en.wikipedia.org/wiki/Permafrost
http://www.bbc.com/earth/story/20170504-there-are-diseases-hidden-in-ice-and-they-are-waking-up
https://www.popsci.com/how-did-anthrax-flare-up-in-siberia/
https://www.popsci.com/waking-up-ancient-viruses-from-melting-frozen-wasteland/
https://www.theguardian.com/world/2016/aug/01/anthrax-outbreak-climate-change-arctic-circle-russia
https://www.vox.com/2017/9/6/16062174/permafrost-melting
https://www.weforum.org/agenda/2017/05/the-deadly-diseases-being-released-by-climate-change/

6 What Is Life?

"... The little things are infinitely the most important"

Sir Arthur Conan Doyle (as Sherlock Holmes)

WHAT IS CAUSING THE DISEASE?

Defining life has been challenging throughout the ages. In many cases, it is difficult to clearly identify an organism, and this is especially true for microscipic organisms. There have been mysterious illnesses and unexplained deaths throughout human history and long before. In the midst of World War I, a curious new enemy emerged. Chinese workers who had traveled from northern China were the first to be diagnosed with the illness, but it soon spread to British, French, and American troops. Hundreds of initially healthy soldiers became ill. Eventually, thousands became ill, and it spread across the globe. This pandemic would eventually kill far more soldiers and civilians than those killed from guns and bombs in the war. The cause turned out to be influenza, what we commonly call "the flu." The virus is not only microscopic, but is hundreds of times smaller than a bacterium, and can only been seen with an electron microscope, which had not yet been invented. So, it was assumed at the time to be a bacterial infection, but was identified as a virus years later. It is transmitted through birds, pigs, and humans (as well as many other animals). More than 80 years later, lung tissue of some of those who died from the disease was excised from the corpses buried in icy soil since the WWI pandemic. The virus was detected in the victims' lungs, although it was nonviable. Nonetheless, the viruses and their nucleic acids were still present after decades in the frozen soil, and were repeatedly frozen and thawed in laboratories, suggesting they might survive in ice for decades, or longer. Bolstering this point, influenza A viruses have been shown to retain viability in lake and pond ice and mud.

In ancient China, Medieval Europe, the early history of North America, and more recently in Asia, and Central and South America millions died from severe dehydration due to diarrhea caused by pathogens in drinking water. Some microbial pathogens can survive all sorts of conditions, including heat, cold, freezing, thawing, and high salt concentrations. One of the major water-borne diarrheal diseases is cholera, caused by the bacterium, *Vibrio cholerae*, which carries a virus that produces the toxin that leads to the disease symptoms. In this case, the partnership between a bacterium and a virus causes the disease. If you have one or the other, no disease symptoms are produced. However, bring the two together, and illness and death may result. Strains of this bacterium can grow in freshwater or seawater, and they have been found in cold water and in ice.

In 1922, Lord Carnarvon and Howard Carter discovered the tomb of Tutankhamen in Egypt. Within a few days, Lord Carnarvon became ill and shortly thereafter was dead. Some termed the death "the curse of the mummy." While his illness may have been unrelated to his exposure to items in the tomb, many others also have become ill while exploring tombs, ancient burial chambers, and caves. Are the dead able to affect us hundreds and even thousands of years after their deaths? No, but the explanations for the illnesses and deaths are just as intriguing. Microorganisms are everywhere, including growing in caves and tombs. Lord Carnarvon probably succumbed to a species of fungus within the genus *Aspergillus*. His death was stated to be from pneumonia, and several species of *Aspergillus* are able to infect lung tissues. Whether Lord Carnavon's infection came from Tutankhamen's tomb, from his shaving kit (he had an infection from shaving), from the water, or from the dust in the air, no one is certain. *Aspergillus* is only one of hundreds of fungi that infect humans. They, like the hundreds of pathogenic bacteria and viruses, can infect humans. Some lead to mild problems, such as diarrhea, while others can cause devastating consequences, including death. Viruses such as HIV, influenza, Ebola, and papilloma viruses have few genes, and are so small that electron microscopes must be used to visualize them. But, they can stop large complex animals (including humans) dead in their tracks. Not only do microbes cause acute diseases, but they are also responsible for chronic disease, such as heart disease, allergies, and cancer. It has been surmised that Charles Darwin died from Chagas disease, caused by a trypanosome (a unicellular eukaryotic protist) that he contracted in South America, while on his voyage aboard the Beagle. The pathogen, which is transmitted by an arthropod vector (i.e., the kissing bug), causes degradation of cardiac muscle, leading to disability, and eventually to death.

Microorganisms are all around us. We breathe hundreds of them in and out with every breath. We drink them in our water, coffee, and soft drinks. They are in the food we eat, many of them survive cooking. They live up to miles deep in the earth, miles deep in the oceans, and miles high in the atmosphere. They are found in clouds, fog, rain, snow, and in the air around us. Most are harmless, and a large number are beneficial. However, there also are many that are pathogenic.

MICROBES IN ICE

Our view of life and where it can survive has changed radically during the past few decades. We know that organisms can survive or even thrive in very extreme environments. Ice cores have been retrieved from glaciers miles below the surface of Greenland and Antarctica (see Fig. 1.2). Permafrost and soils have been excavated from deep in the Earth. Ice, water, and mud from surface lakes and subglacial lakes have been examined. In every case, scientists have recovered living microorganisms, as well as fragments of dead microorganisms, multicellular life forms, and biological molecules. Some microbes survive temperatures of over 200°C (392°F), and some can grow and divide at temperatures over 110°C (230°F). Many live in environments with high pressures, including those on the ocean floor that would crush a man or a submarine (500–1,000 atmospheres of pressure; equivalent to 7,350 to 14,700 lbs/in^2; or 5,160 to 10,330 kg/m^2). Others live in extremely dry conditions, extremely acidic (less than pH 1.0) and alkaline (greater than pH 10) conditions, high salt environments, and some can survive 1.5 million rads of radiation (a lethal dose to humans can be as low as 450 rads). And now we know that a large number of microorganisms survive being frozen in glaciers and permafrost for decades, centuries, millennia, and longer. Viable bacteria and fungi have been recovered from ice more than one million years old, and bacteria have been isolated from permafrost dated to more than 20 million years old. In glacial ice, temperatures can be far below 0°C, and pressures can exceed 7,000 pounds per square inch (5,000 kg/m^2).

Microorganisms characteristic of more temperate latitudes, as well as native polar microflora are present in ice. You might be tempted to think that because the polar regions and the ice are so cold, always at or well below the freezing point of water (0°C = 32°F), the only life found in ice would be that capable of growing at or below the freezing point of water with little or no growth at higher temperatures. While a number of cold-loving bacteria and fungi have been found in ice, many of the microorganisms isolated from ice are those that prefer higher temperatures (they grow best at 20–30°C; or 68–86°F), and most merely tolerate the colder temperatures. They are found in very cold environments, as well as in the oceans, fresh water, garden soil, fresh vegetables, meat, poultry, dairy products, caves, and many other places. Some heat-loving microbes also have been isolated from Siberian permafrost at a depth of 6 feet, in glacial ice, and in Lake Vostok accretion ice. Cold-loving microorganisms may have originated from the polar regions, but the organisms that require higher temperatures probably came from more temperate latitudes by wind action. Wind patterns at the poles have been extensively characterized. It has been shown that winds from lower latitude often reach the poles, carrying with them debris from the entire planet. Microflora characteristic of the atmosphere of more northerly latitudes have been repeatedly detected in the interior of the Antarctic continent.

Recall that ice is an exquisite air-sampler, and therefore, is a repository of life forms present in ancient atmospheres. Scientists have detected viable bacteria and viruses in clouds, so the hypothesis of viable life forms in the atmosphere has been supported by several researchers. It is quite likely that the microbes characteristic of more temperate and tropical latitudes could have been blown into the polar regions and deposited along with snow onto the ice or permafrost surface. If so, they would have become entrapped in the ice and permafrost along with the local flora. They survive in the ice in part because they can tolerate the cold temperatures, or they produce dormant endospores or other resting structures that allow them to endure the severe environment until conditions improve. There are notable exceptions, the thermophilic bacteria found in Lake Vostok. These bacteria may have originated in hydrothermal areas of the lake or in hot rocks below the lake.

How can organisms and their nucleic acids survive for such long periods in the ice? The answer turns out to be more complex than the question of whether or not they survive. It is clear that under certain conditions survival for millennia is common. Survival for hundreds of thousands of years has been demonstrated for many different organisms preserved in a number of ways. Survival for more than several millions of years also has been demonstrated in a few research studies. Beyond about 1–3 million years, however, the number of survivors and the evidence for survival begins to thin out considerably. Some are capable of metabolic activity while entombed in ice. They probably can repair their genomes, as well as other cellular components. It appears that a combination of cold, dehydration, neutral pH, low oxygen, and darkness are sufficient to preserve organisms and their biomolecules for very long periods of time. In many cases, these conditions exist in glacial ice. If an organism can survive dehydration and/or freezing (as can the ones that grow after you thaw out something from your freezer, including bread yeast) then they stand a good chance of being preserved for long periods of time in glaciers.

How do they reach the glaciers? How many are released worldwide? What does this mean in terms of the evolution and evolutionary rates of these organisms? How would release of these organisms through melting of the ice affect the environment, plants, animals, and human populations? Could global warming unleash large numbers of pathogenic organisms entrapped in polar ice? Are some of these organisms so ancient that human populations would not have any immunity to them? Can global cooling slow these processes? These questions, and others have been asked by biologists who study microorganisms in ancient ice. The answers are out there right now. Although all of the answers will eventually be evident, we hope that they will come in the form of careful research studies, rather than in the form of serious disease epidemics.

CLASSIFICATION OF LIFE

All free-living (autonomous) cellular life forms are composed of membrane-bound cells, which carry out the basic life functions of metabolism (the accumulation, conversion, and transformation of nutrients and energy) and reproduction (the ability to replicate and transfer genes to offspring). In very simple terms, cells are biochemical factories capable of converting energy in various ways. Some capture energy directly from the environment (e.g., light energy for photosynthesis in autotrophic organisms), while others eat compounds or other organisms (i.e., for heterotrophic organisms). The energy from these sources is used to synthesize the products necessary for them to grow and reproduce, all within metabolic processes. All cells discovered to date are either bacteria, archaea, or eukaryotes. Often bacteria and archaea are grouped together as "prokaryotes." However, the term "prokaryote" implies that they are cells without nuclei. The point has been made that this is a negative definition, similar to defining humans as being organisms without leaves. Also, one group of bacteria, the Planctomycetes, have double membrane-bound internal organelles that are nucleus-like. Therefore, here we will avoid using the term "prokaryote." Bacteria, archaea, and eukaryotes have all been found frozen in glaciers, permafrost, and other environmental ice, and many were determined to be viable.

All cells evolved from a common ancestor (termed LUCA, last universal common ancestor) that existed on Earth more than 4 billion years ago (Fig. 6.1). Bacteria, archaea, and eukaryotes are not fundamentally different in terms of biochemistry. In fact, archaea diverged early from bacteria, and eukaryotes are essentially endosymbiotic assemblages of an archaea, and one or more bacteria or other eukaryotes. Bacteria are split into Gram positive (monoderms, having a single membrane, most of which take up Gram stain) and Gram negative (diderms, having an inner membrane and an outer membrane, most of which have no reaction with Gram stain). Eukaryotes have a more complicated evolution. The first eukaryote was produced by an endosymbiotic event between an archaeon (which by that time had diverged from bacteria) and an organism with a nucleus-like internal organelle, possibly an ancestor of a planctomycete. These first eukaryotes had no mitochondria. Next, one of these cells formed an endosymbiotic relationship with an alpha-proteobacterium, which became the mitochondria.

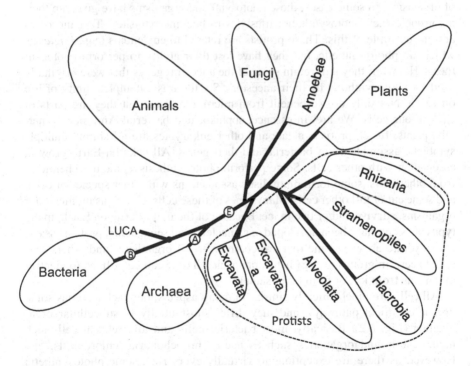

FIGURE 6.1 Phylogenetic (evolutionary) tree showing the relationships among cellular life on Earth. Solid lines connecting the groups of organisms indicate vertical inheritance patterns. The three Empires are indicated by the basal three branches (A—Archaea; B—Bacteria; E—Eukarya). Protists include Excavata a (including trypanosomes—mainly parasitic), Excavata b (including euglenoids—mainly photsynthetic), Alveolata (e.g., dinoflagellates and ciliates), Hacrobia (e.g., cryptomonads and haptophytes), Stramenopiles (e.g., brown algae and diatoms), and Rhizaria (e.g., foraminifera and cercozoa). LUCA is the last universal common ancestor, from which all other organisms originated.

Alphaproteobacteria are capable of performing oxidative phosphorylation. This allowed nearly 20 times as much energy in the form of ATP to be extracted from each glucose molecule, through the use of the enzyme ATPase. In the process, oxygen and hydrogen are joined to form water. [This is the reason that we all need to breathe in oxygen, and breathe out water vapor, along with carbon dioxide, which comes from another part of our metabolism.]

Later still, photosynthetic bacteria were engulfed by some eukaryotic cells and became the chloroplasts. These cells are known as algae, and their descendants are the higher plants. Beyond this, other endosymbiotic events between several types of protists have led to the evolution of other types of protists. This includes water molds, dinoflagellates, trypanosomes, euglenoids, brown algae, ciliates, and many others. In most cases, the endosymbiotic events occurred when a non-photosynthetic protist engulfed a red or green alga (i.e., plant cells), thus becoming photosynthetic. In a few cases, some of these organisms engulfed yet another red or green alga, so they contain genes that originated in several different lines of organisms. In some cases, these photosynthetic organisms have given up their autotrophic (i.e., photosynthetic) lifestyles to become parasites. Trypanosomes are one example of this. These protists are related to euglenoids (e.g., *Euglena*), which are photosynthetic, but they have lost their ability to perform photosynthesis. However, they still retain some of the ancestral genes that were originally used for photosynthesis by their ancestors. So, there is a complete unity of life on Earth. Not only do we benefit from microorganisms, but they are parts of each of our cells. We are, in essence, sophisticated bacteria! And more generally, plants, fungi, protists, algae, and other eukaryotes are all simply multiple symbiotic assemblages of bacteria, and their genes. All cells on Earth grow in association with other cells. Some perform photosynthesis, some use chemicals from other cells, some form mutualistic associations with other species of cells, and some eat or kill other cells. Many types of these cells, as well as multicellular organisms, survive and/or live in ice. Because of the unity of life on Earth, many types of assays have been developed (e.g., nucleic acid sequencing, protein assays, carbohydrate assays), some that are common to all organisms, and others specific to certain groups of organisms, or species. These can be utilized to identify organisms from ice, or from other samples.

All cells are biochemically similar in many aspects, but each also has some unique chemical pathways, and they differ substantially in subcellular organization (Fig. 6.2). Basically, most bacterial cells and all archaeal cells lack membrane-bound organelles such as nuclei, mitochondria, chloroplasts, etc. However, as there are exceptions to virtually every rule, some photosynthetic bacteria do contain membrane-bound vesicles that resemble organelles, and some planctomycetes have internal membrane-bound organelles, including ones in *Gemmata obscuriglobus*, which has double-membrane organelles that resemble nuclei (Figs. 6.1 and 6.2). Eukaryotes include fungi, amoebae, protists, animals, and plants. These organisms consist of cells that possess membrane-bound organelles, including nuclei, which are surrounded by a single membrane that is folded over on itself, and thus appears to be two membranes. Some also

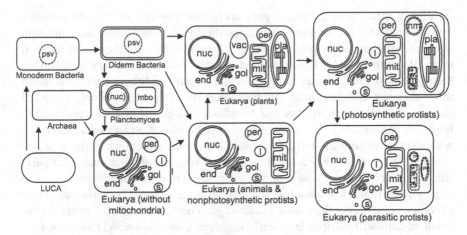

FIGURE 6.2   General cellular organization for bacteria, archaea, and eukaryotes, show-ing the unity of the major groups of organisms. LUCA (last universal common ancestor) had a single membrane and relatively simple processes. This led to the evolution of mono-derm bacteria (single cell membrane) and archaea. The diderm bacteria (two cell mem-branes) evolved from monoderm bacteria, with additional cell complexity, as well as the central biochemical pathways from the monoderms. Photosynthetic bacteria have internal photosynthetic vesicles (psv). A number of endosymbiotic events led to various members of eukaryotes: an archaea and a planctomycete to produce eukaryotes with a nucleus, but no mitochondria; a eukaryote and an aphaproteobacterium (diderm bacterium) to produce mitochondrial eukaryotes; a mitochondrial eukaryote and a photosynthetic cyanobacte-rium (diderm bacterium); several protists (eukaryote) and red and green algae to produce photosynthetic protists. Some of the photosynthetic protists became parasites and lost their photosynthetic functions. Abbreviations: end—endoplasmic reticulum; gol—Golgi apparatus; l—lysosome; mbo—membrane-bound organelle; mit—mitochondrion; nuc—nucleus; (nuc)—nucleus-like organelle; per—peroxisome; pla—plastid (chloroplast or other related organelle); s—secretory vesicle; vac—vacuole.

contain mitochondria and chloroplasts, which also have two surrounding mem-branes. Other organelles are the result of a eukaryote engulfing one or more other eukaryotes. In these cases, the organelles are usually enclosed by more than two membranes. Bacteria were clearly present at least 3.5 billion years ago, while evidence of Archaea are from 3.0 billion years ago. Eukarya were first present by about 2.4 billion years ago. Most were unicellular, but multicellular species had begun to appear as early as 2.1 billion years ago. All descended from a common ancestor that appears to have originated about 4.2 billion years ago. Having a common ancestor means that all life on Earth is related and thus can be represented on one single evolutionary tree, often termed the "tree of life." Virtually every phylum of organisms contains some members that are parasitic and/or pathogenic. In addition to biochemical assays and analyses, morphologi-cal and physiological characteristics can also be used to identify many types of organisms.

Viruses differ from cellular organisms in that they all are obligate parasites (i.e., they need host cells to reproduce), and they have no cell membrane, although many are surrounded by part of their host's membrane once they leave that cell (e.g., HIV, influenza A). Theories about the origins of viruses vary greatly. One theory proposes that viruses were the original forms of life on Earth that led to more elaborate cellular forms. Another states that viruses are simply pieces of genomes of cellular organisms that found ways to get themselves replicated in host cells. Another has shown that viruses can be placed on an evolutionary tree with all other organisms, and that viruses form a coherent branch of that tree. Yet another theory is that they are simply reduced versions of what once were parasitic cellular organisms. This is supported by some of the characteristics found in large viruses that infect amoebae. They are as large as small bacteria, have large genomes (some larger than some bacterial genomes), some of their genes are for standard cellular functions (e.g., cell membrane components, ATPase subunits), and they can be infected by other viruses (called virophages). So, they are right at the cusp between cellular and non-cellular life. Whatever the true evolution of viruses, because they contain nucleic acids, and some are capable of some metabolic functions while inside their host cells, they can be considered living organisms, although much reduced ones. Thus, life is diverse, widespread, complex, and not easily defined.

SOURCES AND ADDITIONAL READINGS

Bidle, K.D., S. Lee, D.R. Marchant, and P.G. Falkowski. 2007. Fossil gene microbes in the oldest ice on Earth. *Proc. Natl. Acad. Sci. USA* 104: 13455-13460.

Castello, J.D., and S.O. Rogers. 2005. *Life in Ancient Ice*. Princeton, NJ: Princeton University Press.

Castello, J.D., D.K. Lakshman, S.M. Tavantzis, S.O. Rogers, G.D. Bachand, R. Jagels, J. Carlisle, and Y.J. Liu. 1995. Detection of infectious tomato mosaic tobamovirus in fog and clouds. *Phytopathology* 85: 1409-1412.

Christner, B.C., J.C. Priscu, A.M. Achberger, C. Barbante, S.P. Carter, K. Christianson, A.B. Michaud, J.A. Mikucki, A.C. Mitchell, M.L. Skidmore, T.J. Vick-Majors, and the WISSARD Science Team. 2014. A microbial ecosystem beneath the West Antarctic ice sheet. *Nature* 512: 310-313.

D'Elia, T., R. Veerapaneni, and S.O. Rogers. 2008. Isolation of microbes from Lake Vostok accretion ice. *Appl. Environ. Microbiol.* 74: 4962-4965.

D'Elia, T, R. Veerapaneni, V. Theraisnathan, and S.O. Rogers. 2009. Isolation of fungi from Lake Vostok accretion ice. *Mycologia* 101: 751-763.

Ma, L., C. Catranis, W.T. Starmer, and S.O. Rogers. 1999. Revival and characterization of fungi from ancient polar ice. *Mycologist* 13: 70-73.

Ma, L., S.O. Rogers, C. Catranis, and W.T. Starmer. 2000. Detection and characterization of ancient fungi entrapped in glacial ice. *Mycologia* 92: 286-295.

Petit, J.-R, J. Jouzel, D. Raynaud, N.I. Barkov, J.-M. Barnola, I. Basile, M. Bender, J. Chappellaz, M. Davis, G. Delaygue, M. Delmotte, V. M. Kotlyakov, M. Legrand, V. Y. Lipenkov, C. Lorius, L. Pépin, C. Ritz, E. Saltzman, and M. Stievnard. 1999. Climate and atmospheric history of the past 420,000 years from the Vostok ice core, Antarctica. *Nature* 399: 429-436.

Rogers, S.O. 2017. *Integrated Molecular Evolution*, 2nd ed. Boca Raton, FL: Taylor & Francis Group.

Rogers, S.O., Y.M. Shtarkman, Z.A. Koçer, R. Edgar, R. Veerapaneni, and T. D'Elia. 2013. Ecology of subglacial Lake Vostok (Antarctica), based on metagenomic/meta-transcriptomic analyses of accretion ice. *Biology* 2: 629-650.

Shoham, D., A. Jahangir, S. Ruenphet, and K. Takehara 2012. Persistence of avian influenza viruses in various artificially frozen environmental water types. Influenza Res. Treat. dx.doi.org/10.1155/2012/912326.

Shtarkman Y.M., Z.A. Koçer, R. Edgar R, R.S. Veerapaneni, T. D'Elia, P.F. Morris, and S.O. Rogers. 2013. Subglacial Lake Vostok (Antarctica) accretion ice contains a diverse set of sequences from aquatic, marine and sediment-inhabiting Bacteria and Eukarya. PLoS ONE 8(7): e67221. doi:10.1371/journal.pone.0067221.

Taubenberger, J.K., A.H. Reid, R.M. Lourens, R. Wang, G. Jin, and T.G. Fanning. 2005. Characterization of the 1918 influenza virus polymerase genes. *Nature* 437: 889–893.

Zhang, G., D. Shoham, D. Gilichinsky, S. Davydov, J.D. Castello, and S.O. Rogers. 2006. Evidence for influenza A virus RNA in Siberian lake ice. *Journal of Virology* 80: 12229–12235.

LINKS

https://en.wikipedia.org/wiki/Spanish_flu

7 Fossils: Marking Time

"If I could keep time in a bottle"

Jim Croce

READING FOSSILS

Fossils are biological materials from the past that are preserved in various matrices (e.g., rocks and amber). As such, microbes and biological materials entrapped in ice can be included. Fossils have intrigued people for thousands of years. When Charles Darwin departed England on December 27, 1831 aboard the H.M.S. Beagle, he was more interested in fossils than finches or tortoises. Even as the voyage, exploration, and collections progressed, he retained his fascination with fossils and sent large collections back to the British Museum. While collecting extant organisms in eastern and western South America, he also collected many fossils from the rock formations. He was interested in the extinct life forms and what they might tell us about past times, which helped him develop his ideas on evolution and natural selection. As he and the Beagle crew dined on Galapagos tortoises (believing that they were not native to the islands, but had been deposited onto the islands earlier to serve as food for the sailors) and as he mounted the later-infamous "Darwin's" finches that he had collected there (most of which he did not initially identify as finches, John Gould at the British Museum was the person who later sorted this out), he carefully cataloged and boxed fossils that he had recently collected in the Andes. Many represented new species, including one that was ultimately named after Darwin (*Mylodon darwini*, an extinct giant sloth). Fossils are a window into the past, and also provide a wealth of information about evolutionary and genetic processes.

Fossils have provided important evidence for many biological and geological fields of study. Fossilized bacteria that are 3.5 billion years old are remarkably complex. They provided evidence that life appeared relatively soon (i.e., within a few hundred million years) after the Earth formed and began to cool. This conclusion has been confirmed by other evidence, which has pushed the origins of life on Earth to at least 4.2 billion years ago. In addition to external and sometimes internal morphology, certain fossils can indicate behavior traits (tubes in mud, walking, mating, foods in gut, feces, etc.). Fossils have also led to some geological discoveries, including the phenomenon of continental drift. Some fossils and rock formations found in western Africa and eastern South America are morphologically identical, indicating that the two continents had once been attached. Other sets of fossils and rock formations have shown the time course of continents joining and splitting through billions of years. Some indicated that dinosaurs, trilobites, and amonites evolved, diversified, dominated, and then went extinct, all at different times, millions of years apart. They have shown us that giant sloths and rhinos once roamed North and South America. This and other evidence lead to the conclusion that the Earth's biosphere changes continually, and sometimes dramatically; and extinction is common. More than 99% of all the species that have ever existed on Earth are extinct. Fossils were part of the evidence to show that at least some of the mass extinctions were caused by impacts of large meteors (asteroids or comets), and other events. The mass extinction that occurred at the end of the Permian period, approximately 252 million years ago, was among the largest, where more than 90% of the species suddenly went extinct.

Fossils have allowed scientists to reconstruct the evolutionary histories for many of the major groups of plants and animals. They have answered important biological and evolutionary questions; e.g., the yet unknown cause of the Cambrian explosion of species, a period about 550 million years ago when all of the multicellular animal body plans arose, many of which have since disappeared. Following this period there was an increase in plant and fungal diversity. This drastic increase in the diversity of organisms on Earth is being studied primarily by using hard fossils. One theory for this rise in diversity is that it was caused by a major ice age, one so cold that much of the Earth was covered with ice, including ice covering the oceans all the way to the equator. This "Snowball Earth" hypothesis states that these large areas of ice created isolated islands and bodies of water, which separated species from one another. Isolation is one of the major causes of speciation. Thus, ice may have had major influences on evolution, and is probably still a major influence in this respect. Another characteristic of the Cambrian explosion is that animals evolved hard body parts, indicating the possible increase in predation and the need to develop protective armor.

FOSSILS IN ICE

The Earth is not static. Large and small changes have occurred in the past and will occur in the future. You walk through a world today that is different from the one that you walked through yesterday, and different still from the one that you will walk through tomorrow. The environment changes during the stages of the day, the season, the millennia, and beyond. Human-caused pollution has been changing the environment for hundreds of years, and these effects continue to increase. Fossil remnants record these changes, leaving clues of the past. However, that past cannot be reconstructed from any individual fossil. It is only by viewing many different fossils and fossil types that we can reconstruct the past. For example, millions of tyrannosaurs walked the Earth during the millions of years of their reign. Although very successful, they all died. Some were covered by mud and debris, became compressed and compacted, eventually becoming fossil remains. Only a small number of these have been unearthed by scientists, while countless others remain buried. The ones that are found become our fragmented windows into the past. Which ones are found and the state of their preservation determine the clarity of that window. In general, the older they are, the more they have been damaged by a number of chemical and physical processes, and thus the fewer that now exist in a usable condition.

The window for microbes is much hazier, because these fossils exhibit few morphological details that can be used to identify species. Most of the time only partial impressions of their outer surfaces remain. There are some bacterial fossils dating to more than 3.5 billion years ago, but only grainy surfaces are evident. Little information as to their evolution or physiology can be determined from these ancient remnants. Most bacterial remains are severely or completely degraded in only weeks to months. Only a precious few are preserved for thousands of years or longer, so it is extremely rare to find some that are millions or billions of years old. However, glaciers and permafrost hold a large number of well-preserved microbes, some of which can actually be revived after thawing. Thus, their physiology, genetics, and morphology can be studied in great detail. Although most of the microbes found so far in ice and permafrost on Earth are less than 500,000 years old, a few have been found in ice and permafrost that are up to 25 million years old. Ice on planets, moons, asteroids, and comets may be much older than this. It will be very interesting when some of these are examined in detail.

LIVING FOSSILS

The hundreds of types of microbes that we inhale into our lungs today are different ones from those that we will inhale tomorrow. Happily, most are not human pathogens, and those that are soon fall prey to our immune systems. Microorganisms grow, reproduce, and die, and they also move in the soil, wind, clouds, fog, water, snow, and ice. Streams, rivers, lakes, oceans, and glaciers are also in constant motion. They carry sand, gravel, rocks, boulders, animals, plants, and microbes. Many of the microbes (and some multicellular organisms) can survive freezing. The conditions at the time of freezing and entrapment of microbes in glaciers dictate the chances for their survival and/or preservation of their remains. Some of the microbes become icy fossils that can come back to life after years, centuries, millennia, or longer of dormancy. The conditions at the ends of the glaciers are important, since they dictate whether the microbes can survive in the environment, or in us. Glaciers form, move, and melt, and upon melting they release their entombed cargo. If you could return to a glacial area 50 years from now, it might be a very different scene. With the current acceleration in glacial melting, the glacier terminus might be kilometers from where it was 50 years ago, or it could be gone. Some glaciers change greatly in only a few months or a few years. Enormous numbers of dead and living fossils are released at the termini of all glaciers, and from other exposed surfaces. The fact that the rate of melting is increasing, and that pathogens are included in the mix of organisms, should be of concern to everyone whether they live near a glacier or not. Ice shelves also can change rapidly. Early in 2002, part of the Larsen ice shelf the size of Rhode Island broke off of the mainland of Antarctica. The pieces broke up, each slowly melting into the sea, releasing all of the organisms the ice held.

Glaciers, permafrost, and ice sheets carry unique types and varieties of fossils, including fragments of plants, animals (sometimes including people), and microbes; shells of diatoms; living microbes; and biological macromolecules (e.g., DNA, RNA, proteins, carbohydrates, and others). Many studies are possible using these fossils, including answering questions about evolution, survival, epidemics, pathology, geological events, and human history (and pre-history). Ice cores are valuable resources because each depth or region represents a specific time in the past. The living and dead fossils encased in ice are truly unique, informative, and fascinating resources for scientists. When fossils are brought up in conversation today, most people think of them as being hard rocky mineralized impressions (or pieces) of extinct organisms. However, the fossils in glaciers are very different, and include those that remain viable, and those that consist primarily of biological remains, without mineralization. We, and others, have been examining these for nearly three decades with some surprising results. Ice holds an extremely large number of unique fossils that can be hundreds to millions of years old, which may include a wide variety of viruses, bacteria, archaea, as well as an assortment of eukaryotic organisms, such as protists, algae, amoebae, higher plants, fungi, and animals.

HOW OLD?

How long can organisms remain viable? In past centuries, extracts of ancient organisms were thought to confer longevity to those who ate them. Others (like us) have an intellectual curiosity about why and how organisms survive for long periods. There are obvious advantages to this. Some people have had themselves frozen, in the hope that decades from now they can be brought back to life and cured of any diseases that they had, including old age. But, currently, humans cannot be thawed and brought back to life. However, lotus seeds remain viable for at least several centuries, and one 2,000-year-old lotus seed germinated to grow into a new plant. Estimates on the longevity of bacterial spores ranges from 1,000 to 140,000,000 years. However, in a population of bacterial cells, it has been demonstrated that the number of surviving cells gradually decreases with time. Studies on deep mud cores provide the evidence for survival of bacterial spores (the hard-resting stage for certain groups of bacteria) in excess of 1,000 years old. Viable microbes have been isolated from centuries-old herbarium specimens, thermophilic bacteria have been isolated from ocean basin sediment cores estimated to be 5,800 years old, and bacteria have been recovered from archaeological sites thousands of years old. Various investigators have isolated viable bacteria from ice cores at Antarctica, some of which were deposited more than 400,000 years ago, from the Himalayas in 750,000-year-old ice, from 2.7 to 8.0 million-year-old ice, and from Siberian permafrost up to (and possibly older than) 8 million years old.

As early as the eighteenth century, there were reports of finding microbes in glaciers. In the nineteenth century, a few more such reports were published. In the early decades of the twentieth century, several controversial papers were published that reported the revival of microorganisms from ancient rocks and coal deposits hundreds of millions of years old. Then, in the 1960s, Dombrowski claimed revival of bacteria from 650 million-year-old Precambrian salt deposits. These reports were ridiculed by the scientific community at the time, and the results were attributed to modern contamination caused by sloppy technique. The disbelief and ridicule by the scientific community stifled any further research into the question of ancient life for many years because scientists feared their reputations would be ruined if they published such work. Often, criticism can become biting and personal, and may end careers. The scientific community is extremely conservative and dogmatic when new results challenge conventional wisdom. All reports must be independently confirmed before scientists generally will accept controversial findings. As time passed, isolations of microorganisms from a variety of ancient substrates including the sea floor, rocks, amber, salt deposits, mastodon intestines, old tins of meat, woolly mammoth nostrils, mummified remains of plants and animals, archaeological sites, museum specimens, as well as permafrost and ancient glacial ice, were reported by many researchers, spanning several decades. A database of ancient microorganisms revived after anabiosis for more than 50 years has been created for such reports. It contains a list of thousands of microorganisms recovered from hundreds of separate sites, and reports from dozens of independent research labs. Similarly, the number of microbes isolated from glacial ice now exceeds several thousand, from dozens of sites at a variety of depths (and ages), and reported by about a dozen research labs. Although it is possible that some resulted from contamination, the majority are likely to be accurate accounts of the revival of ancient organisms. Thus, it is apparent that the revival of ancient microorganisms is far more common than most scientists previously suspected.

Theoretical expectations on the longevity of bacterial spores place the limit for recovering viable bacterial spores at about 200,000 years. Clearly, viable organisms have been discovered that extend this at least to over 1 million, and perhaps over 8 million. Reports of the isolation, growth, and DNA characterization of 25 to 40 million-year-old bacteria from amber, and 250 million-year-old bacteria from salt crystals have pushed the theoretical limit to well beyond 250 million years. Empirically based mutation rates that result in detrimental effects on spore viability have been used to estimate the half-life of *Bacillus subtilis* at 7000 years. Assuming an exponential death rate, a large population of viable spores would be detectable after several hundred thousand and possibly for several million years. This is one of the most common bacteria (which incidentally is related to *B. anthracis*, the causative agent of anthrax) that we have isolated from ice cores up to 400,000 years old. These expectations for longevity assume that the immediate environment protects the spores from ultraviolet irradiation, oxidation, and chemical damage common with hydrated nucleic acids. Ice accomplishes all or most of these (depending on the location of the ice), and thus may be one of the best preservatives of microbes and biological molecules (see Fig. 3.2). From all indications, it may be able to protect organisms and biological molecules for millions, and possibly over a billion, years. Additionally, it has been demonstrated that organisms exposed to below freezing temperatures for extended times continue to metabolize at low levels, allowing them to repair any damage to their cells and biological molecules (including DNA), thus extending the theoretical limits of entombment in the ice.

Some of the discoveries we discuss above were initially unbelievable, including the survival of multicellular organisms in ice. In 2018, three hundred nematodes were melted from Siberian permafrost. Two of them were still alive, and began moving and eating. One was from a 32,000-year-old section of permafrost, and the other from a 42,000-year-old piece of permafrost. Quite a Rip Van Winkle event. This followed a research project in which a viable virus that infects amoebae was revived from 30,000-year-old permafrost. This discovery was important, because viruses are incapable of metabolic activity unless they have infected a host cell, which implies that either the host cell was still alive, or that the nucleic acids of the virus were capable of long-term longevity in ice. In 2007, a bacterium was isolated from a piece of 8-million-year-old ice. The isolation of a living microbe from ice that is millions of years old is theoretically impossible (or at least exceedingly unlikely) according to some scientists, and is probably impossible to believe for many non-scientists. However, the majority of the reports of viable microbes in ice have been corroborated in one way or another. Microbes from ancient ice have been grown in laboratories around the world over the past several decades. These living fossils had been frozen in glaciers up to 1 million years and permafrost for up to 25 million years. The researchers (including our labs) have grown and identified thousands of these organisms. Many look very much like extant organisms that have been described previously by scientists, including some similar to those that grow on your shower curtains and bread. Others are unique to science and are not closely allied with any currently described species. Do they represent extinct organisms, or are they simply poorly studied? In many cases, no one knows.

While many of the microbes can be revived, others defy all attempts to culture them. Whether they are dead or living is often a question that cannot be answered. If an organism fails to grow under laboratory conditions, it might be dead or we may simply not know the optimal conditions for its growth. Many of these may be new to science, and many may require very unique growing conditions. Thus, the number of organisms that are culturable in the lab are only a small proportion (estimates range from 0.1% to 1.0%) of the total number of viable microbes in the ancient ice. Therefore, the number of viable microbes actually in those ice samples would range from 500,000 to 5,000,000. To date, thousands have been revived from ancient ice samples around the world. Microbial and biomolecular fossils are now giving us yet another perspective on the past. From the research of various studies (including ours) it has been calculated that hundreds to millions of living microbes are in each liter of glacial ice meltwater, dependent on the location and age of the ice. In addition, there are many times more fossils of dead organisms in the meltwater. They, too, can be studied to search for clues to various scientific puzzles by analyzing the remnants of the biological molecules. Taken as a whole, there exists an enormous repository of living and dead microbial fossils.

ANCIENT BIOMOLECULES

There are still other types of fossils that are especially abundant and well preserved in ice and permafrost, and that yield detailed information of microbial life. Ancient biological molecules, including nucleic acids, amino acids, proteins, polysaccharides, and others are present in most, if not all, glacial ice and permafrost core samples. While rocky fossils can be studied by comparative morphological methods, biomolecules can indicate many other aspects of the fossil. Some can be used to determine the age of a specimen. For example, amino acids in living organisms are in a particular class called L-amino acids. The L stands for levorotatory (literally, left rotating), since the molecules rotate polarized light to the left. However, the amino acids can change over time, such that some change to D-amino acids. The D stands for dextrorotatory (right rotating), since it rotates polarized light to the right, because the arrangement of the atoms in this molecule is the mirror image of that in the L-amino acids. Therefore, if the organism dies its L-amino acids within its proteins cannot be replaced as they slowly change into the D-amino acid forms. The rate of this change (called racemization) is known. Because of these changes, the specimen can be dated according to the relative proportion of L-amino acids to D-amino acids.

Most research to date has been on nucleic acids (DNA and RNA), primarily DNA. Preservation of the DNA is primarily dependent on the limitation of free water (i.e., dehydration), oxidizers (e.g., oxygen), acids, light, and heat. Nucleic acids can also be preserved for long periods of time in ice because the cold temperatures and the scarcity of liquid water protects them from degradative processes (see Fig. 3.2). Ice protects because the water is in solid form, with little "free" water. Water is the primary solvent in cells that moves chemicals around, in, and between them. It also coats all biomolecules, becoming integral parts of those molecules. Without the water, the chemicals do not move and the biomolecules do not function. Therefore, degradative and metabolic processes are greatly slowed. The cold temperatures also slow down degradative and metabolic processes. The ice also limits the amount and effects of gases, such as oxygen, which is a powerful oxidant of biomolecules. When compared to other environments, ice is similar to salt and amber for preserving DNA. However, the isolation of microbes and DNA from salt and amber have not been corroborated in most cases, while the results from similar studies with organisms and DNA from ancient ice have been corroborated many times. Ice might be one of the best preservatives of nucleic acids and probably most other biomolecules. Our work has focused on ancient DNA and RNA, but we also use classical microscopic, cultural, and physiological techniques to study these fossils. Morphology and culturing can identify species in some cases. Physiology can indicate taxonomic affiliation as well, but can also indicate the growth habit of the organism. DNA can be used for all of these in addition to being the primary measure of genetics and evolutionary history. Thus, it is extremely fortuitous that ice is such a good preservative of DNA, as well as RNA.

Nucleic acids also have the potential to indicate a great deal about ancient biodiversity, environments, weather, epidemics, evolutionary processes, geological events, human migrations, etc. When preservation is good, the nucleic acids can be used for sequence and evolutionary analyses. Researchers (including us) have determined and compared DNA sequences from these fossils with sequences from similar contemporary examples of similar organisms. Some of the organisms appear to be unchanged from current members of the same species, some exhibit limited changes, and some change a great deal. These results were previously predicted. However, we have also discovered some very unique organisms and sequences. The most interesting are those that appear to be very different from any that have been previously described. For example, in our metagenomic studies (a method where all of the nucleic acids in a sample are sequenced and then compared with the national and international sequence databases), approximately 50% of the sequences obtained are unlike any in the databases. Of the 50% that do resemble sequences in the existing databases, about 50% of those (25% of the total) are comparable to known species and known genes. The other 50% (again, 25% of the total) are similar to sequences that other scientists have reported, but the species are unknown, in that they have never been described or investigated by anyone else. They are likely species unknown to science, indicating that they are newly discovered species. Some of them may be extinct. Of course, if they are alive in the ice, they are not exactly extinct, but merely have disappeared from the biosphere for a while. This is similar to saying that "the reports of my death have been greatly exaggerated." The organisms were extinct for a while. Nevertheless, it might be worthwhile to try to identify these unknown microbes.

PROBLEMS WITH DEGRADATION

Prior to the reports that ancient nucleic acids were preserved in some fossils, only contemporary sequences could be compared. Therefore, on an evolutionary tree, only the tips of the branches could be sampled, and the branches themselves had to be inferred. This is true for all such trees, even the ones that were determined using sequences from microbes. The so-called "ancient" organisms in these studies are as contemporary as you and I. One can go deeper into the tree to sample from ancient branches using ancient preserved DNA. Therefore, theoretically more accurate evolutionary trees can be constructed. However, DNA and RNA will degrade over time unless there is some metabolic repair in action. The bases (the A's, G's, C's, and T's) may fall off. The nitrogen-containing groups frequently are degraded, and often the backbone of the DNA is cleaved. When attempts are made to sequence degraded DNA, often because the enzyme that is used (i.e., DNA polymerase) fails to recognize the severely degraded template DNA, and cannot read and copy it. Or, inaccurate sequences are obtained. For example, if a base falls off the DNA, any base may replace this in the sequence. Therefore, when these experiments are performed, they have to be repeated several times, sequencing is performed in both directions, and a number of controls have to be used to ensure the accuracy of the sequence. Most Next Generation Sequencing processes normally sequence the same regions over and over again, and then come up with a consensus sequence for each region that has been sequenced. That is, a region is sequenced multiple times, and these copies are lined up. If there are 20 aligned copies, and an "A" appears at one position in 19 out of the 20 copies, then the "A" is used in that position in the consensus sequence. However, for ancient and degraded DNA, this is no easy task, because the DNA is frequently in pieces and there is very little of it. If the "A" is only in 10 of the 20 copies, it is uncertain whether the "A" should be used in that position of the consensus sequence. Many researchers have avoided this field of study for precisely this reason. Nonetheless, research has proceeded and the researchers have made great achievements. The phrases "Ice Biology," "Ancient Organisms in Ice," "Genome Recycling," "Astrobiology," "Cryobiology," and others have appeared with increasing frequency in the literature, at scientific meetings, and in the popular media. This is a good sign for this area of biology.

REAL OR IMAGINED?

Contamination is another problem when working with DNA and RNA (as well as rare organisms) isolated from ancient materials because the ancient DNA is almost always degraded to some extent, and any contemporary nucleic acids will be preferentially amplified by PCR, since they will be completely intact (or nearly so). Therefore, extreme care must be taken when working with these ancient samples. Contamination has been a major problem in several instances. For example, in the late 1980's and early 1990's several publications appeared in major scientific journals reporting DNA sequences and viable organisms isolated and characterized from fossils and amber. However, after several laboratories could not replicate the results, skepticism grew. In a few cases it was experimentally determined that modern contaminating DNA and organisms were included with the specimens used and reported in the publications. Thus, the results were doubted. Most notable was the report of a chloroplast sequence from a 20-million-year-old plant fossil, and a viable bacterium purportedly isolated from the gut of a termite encased in a 140-million-year-old piece of amber. Alas, both have since been refuted, and attempts to replicate the results have all failed. Amino acid racemization assays have shown that the magnolia leaf sample contained modern contaminating organisms. There are still several laboratories around the world studying ancient DNA, but most focus on mummified human, plant, and fungal specimens, and exclude work with rocky fossils, salt deposits, and amber. Members of the ancient ice community, with past problems in mind, have been exceedingly careful in their scientific approaches to isolation and characterization of the encased organisms. Intense scrutiny on the elimination of contaminants and inclusion of multiple controls have been vital to the success of this field of study.

SOURCES AND ADDITIONAL READINGS

Bidle, K.D., S. Lee, D.R. Marchant, and P.G. Falkowski. 2007. Fossil gene microbes in the oldest ice on Earth. *Proc. Natl. Acad. Sci. USA* 104: 13455-13460.

Desmond, A., and J. Moore. 1991. Darwin, The Life of a Tormented Evolutionist. New York: W. W. Norton & Company.

Golenberg, E.M., D. E. Giannasi, M.T. Clegg C.J. Smiley, M. Durbin, D. Henderson, and G. Zurawski. 1990. Chloroplast DNA sequence from a Miocene *Magnolia* species. Nature 344: 656-658.

Pan Terra, Inc., 2003. A Correlated History of the Earth. Hill City, SD: Pan Terra.

Rogers, S.O. 2017. Integrated Molecular Evolution, 2nd ed. Boca Raton, FL: Taylor & Francis Group.

Rogers, S.O. 1994. Phylogenetic and taxonomic information from herbarium and mummified DNA. In: R.P. Adams, J. Miller, E. Golenberg, and J.E. Adams (eds.) Conservation of Plant Genes II: Utilization of Ancient and Modern DNA. Missouri Botanical Gardens Press, St. Louis, pp. 47-67.

Rogers, S.O. and A.J. Bendich, 1985. Extraction of DNA from milligram amounts of fresh, herbarium and mummified plant tissues. *Plant Mol. Biol.* 5: 69-76.

Rogers, S.O., L.-J. Ma, Y. Zhao, V. Theraisnathan, S.-G. Shin, G. Zhang, C.M. Catranis, W.T. Starmer, and J.D. Castello. 2005. Recommendations for elimination of contaminants and authentication of isolates in ancient ice cores. In: Castello, J.D., and S.O. Rogers (eds.) *Life in Ancient Ice.* Princeton NJ: Princeton University Press, pp. 5-21.

Rogers, S.O., Y.M. Shtarkman, Z.A. Koçer, R. Edgar, R. Veerapaneni, and T. D'Elia. 2013. Ecology of subglacial Lake Vostok (Antarctica), based on metagenomic/meta transcriptomic analyses of accretion ice. *Biology* 2: 629-650.

Rogers, S.O., V. Theraisnathan, L.-J. Ma, Y. Zhao, G. Zhang, S.-G. Shin, J.D. Castello, and W.T. Starmer. 2004. Comparisons of protocols to decontaminate environmental ice samples for biological and molecular examinations. *Appl. Environ. Microbiol.* 70: 2540-44.

Shtarkman Y.M., Z.A. Koçer, R. Edgar, R.S. Veerapaneni, T. D'Elia, P.F. Morris, and S.O. Rogers. 2013. Subglacial Lake Vostok (Antarctica) accretion ice contains a diverse set of sequences from aquatic, marine and sediment-inhabiting Bacteria and Eukarya. PLoS ONE 8(7): e67221. doi:10.1371/journal.pone.0067221.

LINKS

https://en.wikipedia.org/wiki/Ancient_DNA
https://en.wikipedia.org/wiki/Larsen_Ice_Shelf
https://www.livescience.com/63187-siberian-permafrost-worms-revive.html
https://www.weforum.org/agenda/2017/05/the-deadly-diseases-being-released-by-cli mate-change/

8 Walking into the Past

"If you should go skating on the thin ice of modern life, . . . don't be surprised when a crack in the ice appears under your feet"

Roger Waters

OBTAINING THE CORE SECTIONS

The first step in the process of obtaining and studying ice cores is to travel to many potential drilling sites. Each site is mapped and evaluated, often by using data obtained from ground-penetrating radar, digging snow pits, and drilling shallow cores. From the data, the depth to bedrock can be measured, and the average annual ice accumulation rate calculated. Also, the evenness of the layers can be evaluated. If the ice has been deposited on, or is moving over, a ridge or uneven rock, the layers may be ragged, twisted, or completely disorganized. Conversely, ice that collects in an ice dome can have very even ice layers. The radar can also indicate whether any body of water (e.g., a subglacial lake) exists under the ice. Once the results are evaluated, they usually are compared with results from other possible drill sites, and decisions are made as to which of the sites will be presented in a proposal to a granting agency. These projects are usually very expensive to undertake. If funding is secured, a complete plan is finalized, including acquisition of personnel, equipment, and supplies, and afterwards, travel and construction of living and work units at the drilling site. This often takes years of preparation and involves travel to relatively inaccessible and often dangerous places. In addition, the logistics of getting people, equipment, and supplies to the site can be time-consuming, difficult, and expensive.

Once the personnel, drilling rig, and housing issues are settled, drilling begins. Drilling of shallow cores can be relatively simple, and so-called "dry drilling," that is drilling without the use of fluids, may be utilized. When drilling involves reaching to hundreds and thousands of meters deep, other issues come into play. One is time. It takes more time to drill each core section as the depth increases, because the drill bit has to be lowered to cut out the next section, and then raised to remove the core sample. Another major issue is pressure. The deeper one goes into the ice, the greater the pressure. When the pressure is released, by drilling and bringing the core to the surface, the release of pressure can cause some of the core sections to crack, split, and sometimes explode. Each of these reduces the amount of usable ice for study. Utilization of drilling fluids (kerosene, diesel fuel, ethylene glycol, hot water, or others) alleviates some of the problems with pressure, such that less cracking and splitting occur, primarily because the release

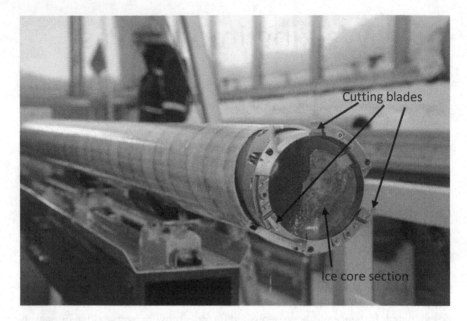

FIGURE 8.1 Ice core drill. The core (1–3 m) is collected inside the lower portion of the drill, which is formed as the cutting blades cut through the ice. Most of the upper portion of the drill consists of motors and computers.

[*Photo credit*: Peter Rejcek and the National Science Foundation]

of pressure is slower when fluids surround the core sections. Another problem is movement and refreezing. If drilling is within a glacier, the glacier is moving, which can cause unequal movement of layers, which can trap the drill. Another thing that can trap a drill is partial or complete refreezing of parts of the borehole. In these instances, drilling fluids are required to keep the borehole open for the drill. The drills are large, complex, and expensive, so having one stuck in the hole is a major issue. The larger drills have a cutting unit with an internal cavity where the cores are collected (Fig. 8.1), but above this, they have motors for the drill, onboard computers to control the drill, receive signals from the surface and relay information from the drill to the surface operators. This is all connected by a large cable that lowers and raises the drill. Interior to the thick cables are wires that connect the drill and computer with the computer and controller at the surface. If a drill and cable are lost, it is a very major expense.

STORAGE

After the core sections are brought to the surface (usually about 1–3 meters each), they are labeled, recorded, and stored at the drill site. Some are later transferred into Styrofoam boxes filled with dry ice, packed into freezers on a ship, and transported to a storage facility. Also, some of the ice is stored at each drill site to ensure that duplicates of the ice core sections are retained in case the adjacent sections of the core are damaged during shipping. In the US, the National Science Foundation Ice Core Laboratory (NSF-ICL) in Lakewood, Colorado is the central facility where most of the ice is stored (Fig. 8.2). However, there are other locations around the country and the world where various collections of ice are housed. Individual researchers and research groups then may request specific core sections (Fig. 8.3) from these repositories, which are stored in the researchers' labs (Fig. 8.4). Some cores are sampled heavily, while others receive few requests, primarily due to the research interests, the condition of the core, the value of the core (where it is from, its depth, historical events, etc.), and other considerations. Each core is unique in many ways, and thus valuable. They therefore are provided sparingly in many cases. Great care is taken to protect the core sections during drilling, shipping, storage, and ultimate study. Even minute cracks can allow the introduction of modern contaminating microbes and chemicals. If this happens, then it becomes impossible to study the ancient organisms, because there is no reliable way to determine which are the old organisms and which are the modern ones. Decontaminating the outside surfaces of the core sections, while protecting the microbes inside, is one of the most difficult problems with this type of research.

The highest densities of ice cores and the deepest cores have been drilled in Greenland and Antarctica (see Fig. 1.2). The deepest core from Greenland is from

FIGURE 8.2 Ice cores inside stainless steel canisters at the National Science Foundation Ice Core Laboratory (NSF-ICL), Lakewood, Colorado. The room is kept at a constant -36°C (-33°F).

[*Photo credit*: NSF-ICL]

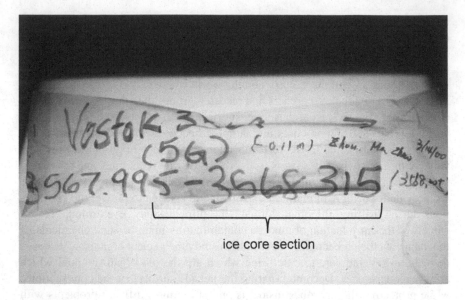

FIGURE 8.3 A single core section, sealed in plastic and kept at a constant -20°C (-4°F) from the Vostok 5G ice core. This core section is from 3567.995 to 3568.315 meters in depth.

a site called GISP2 (for Greenland Ice Sheet Project, site 2) that is over 3,000 meters (9,842 feet) deep (Fig. 8.5). The ice at that depth was deposited almost 240,000 years ago, about 40,000 years before the emergence of *Homo sapiens*. The Vostok, Antarctica core is over 3,700 meters (12,140 feet) deep. The ice at the 3,310 m level of the glacial ice is about 414,000 years old (Fig. 8.5). However, the next 228 m consists of basal ice that is twisted, tumbled, and crushed (due to scraping against the bedrock) so that it cannot be accurately dated. Some estimates suggest that some of the basal ice is more than one million years old. Below that, there is younger ice that originates from subglacial Lake Vostok water. This is called accretion ice and is discussed in detail in a separate chapter. Some older ice on Earth approaches 8,000,000 years old. Some permafrost is approximately 25 million years old.

WHAT TO LOOK FOR

When seeking and characterizing life in ice on the Earth and elsewhere it is difficult to know the types of organisms that might be present. What organisms do you look for? If bacteria or archaea are chosen, which ones will be sought? What are their biological and chemical signatures? Since it is likely that organisms have been blasted off of the Earth by meteor impacts and volcanoes, there are probably organisms that originated on Earth that have fallen onto other bodies in the Solar System, and vice versa. Whether they or their remnants survived the journey is

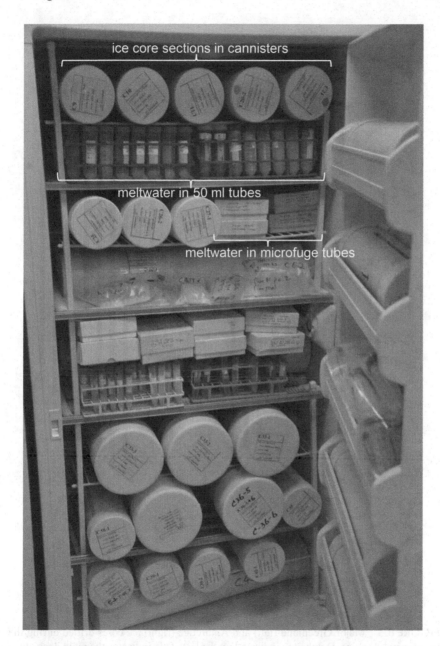

FIGURE 8.4 Ice core sections and meltwater in our lab. A variety of ice core sections from Greenland and Antarctica, as well as tubes with meltwater from other ice core sections. All are stored at a constant -20°C (-4°F).

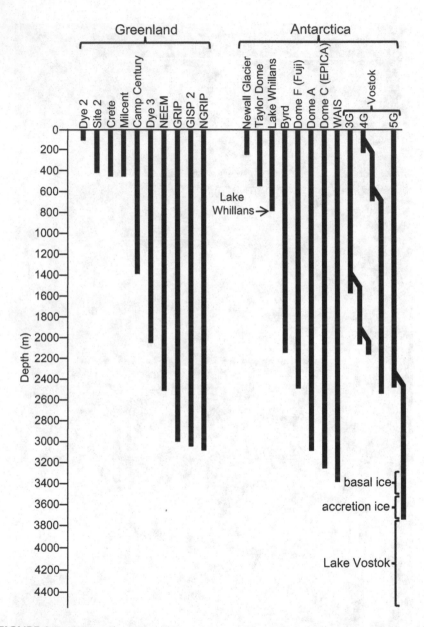

FIGURE 8.5 Major Greenland (left) and Antarctica (right) ice cores drilled during the past several decades. Depths are indicated on the left. This indicates the main deep cores at each site. Additional shallow cores were drilled at some of the sites. For the Lake Whillans project, a hot water drill was used to reach the lake, and then a sterilized drill was used to collect water and sediment samples. For the Vostok cores, the drills encountered obstructions or became stuck several times, so deviation drilling to proceed around the problem spots was used multiple times in the three cores (3G, 4G, and 5G) that were drilled.

an open question. However, microbes have been found alive miles into the atmosphere, and some have survived a trip to the moon and back. If microbes have been able to survive space travel to other planets, how would scientists determine whether or not they originated on that planet, on Earth, or elsewhere? In an analogous way, if organisms evolved first elsewhere in the Solar System or beyond, this could have been the inoculum for the origination of life on Earth (the "Arrhenius" hypothesis). If life is found elsewhere in the Solar System or in the Universe that has similar pathways for replication, transcription, and translation, it is likely that it has an ancestor in common with life on Earth. Chemical signatures for life can also be measured, including changes in methane levels, which are normally due to biological activity.

When one speaks of various ages of organisms on Earth, the common one that first comes to mind is usually "The Age of Dinosaurs," or "The Age of Man." However, dinosaurs, as a group, existed for less than 200 million years. From 4 billion years ago until 2 billion years ago, only unicellular organisms existed on Earth, which has been termed "The Age of Microbes." Some consider that currently we are in "The Age of Mammals," since mammals are physically some of the most complex organisms on Earth. But, mammals have only existed for 200 million years. Others favor the idea that we are in the "Age of Insects," based on number of species, although their numbers are currently decreasing rapidly. However, if we look around our world today, we find that we are still in "The Age of Microbes." If the total biomass of all of the microbes on Earth is calculated, it far surpasses the biomass of the next largest group (higher plants). On the basis of the number of cells, the separation is even greater. Thus, microbes still dominate the Earth's biosphere. In terms of amount of time on Earth, and the variety of environmental conditions under which they live, they surpass all other groups of organisms. Bacteria (as we now define them) have existed on Earth for at least 3.8 billion years, and their ancestors predated them by several hundred million years. The first eukaryotes (all single-celled microbes) appeared about 2.4 billion years ago. Multicellular life forms began to appear about 2 billion years ago, and then greatly increase starting about 600 million years ago. Dinosaurs arose about 250 million years ago, and then disappeared 65 million years ago. The first human ancestors (hominins) originated about 5 million years ago, and the first of the *Homo* species appeared about 2.5 million years ago. Finally, our species originated only 200,000 years ago in central Africa. So, the microbes can trace their lineages back at least 3,800,000,000 years, while human lineages go back only 2,500,000 years for the human-like ancestors, and 200,000 years for modern humans. Therefore, on the basis of time on Earth, biomass, and many other measures, the microbes have been and continue to be the predominant group of organisms on the Earth. Microbes live in hot pools, frigid water, ice, deep oceans, deserts, deep in the Earth, high in the atmosphere, on plants, in plants, on your skin, and in your guts. In other words, they are just about everywhere. Pick almost any location and conditions, and the likelihood of finding microbes is high, or at least higher than finding other types of organisms. This is why researchers working with ice will continue to study primarily microbes. This is also why the search for life on other planets is focused on microbial life.

Isolating and characterizing organisms from ice is difficult. Many of the organisms are completely unknown to science, while others can be clearly identified. Often it is difficult to understand exactly what they are, how they became entrapped in the ice, whether they are alive, and whether they are beneficial, innocuous, or dangerous. A major problem in attempting to isolate and identify them or their nucleic acids is that it is almost impossible to keep the outside of the ice cores free from modern contaminants. As soon as the ice drill touches the ice, modern contaminants are introduced. When the core reaches the surface, airborne organisms stick to the core surfaces. Workers and researchers then handle the cores, and breathe on them, introducing other communities of microbes. These are only some of the potential sources of contamination. Nevertheless, one has to ensure that modern contaminants are not introduced into the interior of the ice core. Core sections must be used that have no cracks where microbes could have entered the interior of the core. Then, the outside of the ice core section must be completely cleaned (decontaminated) of all of the external microbes and nucleic acids. Once this is done, the interior of the ice core section can be melted, but only at low temperatures, because many of the organisms are killed if the temperature rises above about 15°C (59°F).

DECONTAMINATION

Contaminating microbes and chemicals are major concerns. Contamination of the drilling site has been a major issue. However, more care is being taken in the present and planned projects. In ice, there are two parts to this process, and concurrently two major problems. First, the ice core sections are almost always coated in microbes that have come from the drilling process, drilling fluids, people handling the core sections, transport, and the air at the drill site, transfer and storage sites, and research labs. Some of the contaminating organisms are difficult to kill and remove. The second major problem is to thaw and isolate the microbes and/or the biological molecules (e.g., DNA and RNA) without killing the microbes and destroying the biomolecules. In most ice samples, the concentrations of microbes and biomolecules are much lower than for other samples, such as seawater and lake water. The concentrations of contaminants on the outer ice surface of the ice samples are usually higher than the concentrations of organisms within the ice. Also, many of the organisms within the ice are psychrophiles, which can only survive temperatures below 10–15°C, or psychrotolerant organisms, which can grow slowly at low temperatures, but usually grow best between 10 and 25°C. Therefore, melting must be performed at low temperatures, and the meltwater must be kept cold. These organisms often have certain nutrient requirements, so if culturing is attempted, educated guesses must be made when choosing growth media and conditions.

Several decontamination methods have been developed throughout the past several decades. The first was to use water to simply rinse the ice core. This was a poor method, because most water contains some (usually many) suspended microbes. Thus, contamination is a certainty. In the early 1930s, glacial ice and permafrost were tested for bacteria by surface disinfecting the outer surfaces of the samples with a flame and plating the meltwater on suitable bacterial growth media. This relatively crude method for detecting microorganisms in ice remained basically unchanged for 50 years. This will kill many microbes that come into contact with the flame, but much of the ice will immediately melt. Melting absorbs a great deal of heat and can insulate some contaminating microbes from the heat. Thus, this method is inadequate for decontamination. Nevertheless, yeasts and bacteria were isolated from both glacial ice and permafrost using this method. A greater number of bacteria were isolated from permafrost than from glacial ice, and more bacteria were isolated when the media in which the meltwater was cultured was incubated at cold rather than warm temperatures. Since the initial reports, many scientists have reported the isolation of viable microorganisms from recent and ancient ice, as well as from Siberian permafrost that was millions of years old. Until recently, the methods used to detect microorganisms in ice were exactly as described above. But in the past 20–30 years, cultural methods have been refined, decontamination protocols improved, and molecular methods used to amplify microbial DNA directly from meltwater.

One set of methods uses ethanol to treat the outside of the core (Fig. 8.6). However, while ethanol kills many types of organisms, it will not kill all microbes. It is not a sterilant, so it cannot assure the elimination of all outer contaminants. A simpler method simply soaks the core section in sterile distilled water, which is poured off, and then the core section is melted. This method has been demonstrated to fail in removing outer contaminants. One method uses sterilized scraping tools to ablate off the outer surfaces of the ice core sections. This has been shown to be ineffective in removing all outer contaminants. Another method utilizes sterilized drills and saw blades to cut out the central portions of the ice core. However, some of the outer contaminants can be introduced into the central core section by this method. Ultraviolet (UV) light also has been used as a decontaminant, because it can damage nucleic acids, proteins, and other biological molecules. However, UV-irradiation can travel through ice, so that only parts of the outer surfaces of the ice can be decontaminated, otherwise interior organisms also will be killed. Plus, not all organisms are killed by UV-irradiation. Many species of bacteria, fungi, and other organisms are resistant to UV-irradiation. We tested several species of fungi isolated from the ice cores, as well as their nucleic acids, and found that the vast majority were killed after 5–10 minutes of exposure to a germicidal UV-lamp. However, one fungus, *Ulocladium atrum*, was unaffected by the same treatment. When we tried a 15-minute exposure, the fungus still grew normally. It grew as if it had never been exposed to the UV source. Even 30- to 60-minute treatments produced the same effects. The fungus appeared to be completely resistant to ultraviolet light. In fact, when we treated it for 60 minutes with the germicidal UV-irradiation, it seemed to stimulate its growth! Yet another method uses a hot metal cone that is pushed up the core from the bottom, following an ethanol pre-treatment. The melted ice is then collected below.

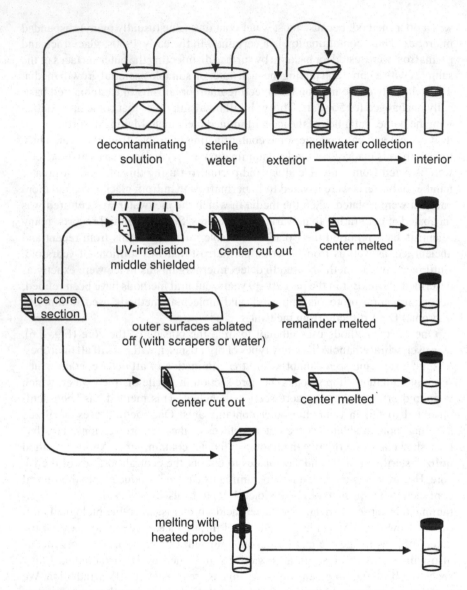

FIGURE 8.6 General decontamination and melting methods. Top line: Ice core section is immersed in a decontaminating solution (e.g., Clorox, ethanol), and then rinsed in large amounts of sterile reverse osmosis water. The section can then be melted in fractions (i.e., shells), or as one sample. Second line: After treating the surface with a decontaminating solution, the center of the core section is covered with aluminum foil or paper and treated with UV-irradiation. The center is then cut out and melted. Third line: The outer surfaces are ablated away, either with sterilized scraping tools (e.g., scalpels, chisels), or with water washes. Fourth line: After treating the outer surfaces with a decontaminant, the center is cut out and melted. Fifth line: After treating the outer surfaces (as above), a heated probe is forced up through the ice core section and the meltwater is collected.

This method has been used in many studies and appears to be mostly effective. However, as mentioned above, many of the organisms are psychrophiles or psychrotolerant, and therefore, some of these may be killed by the high temperatures used in this method.

Because none of the methods assured complete removal of all outer contaminating organisms and biological molecules, we looked for another decontaminant. There were several possibilities: ethanol, hydrogen peroxide, methyl bromide, chlorine, sodium hypochlorite, and others. It turned out that many organisms are unaffected by ethanol. Methyl bromide is difficult to control, because it is a gas at room temperature and is toxic to humans. Hydrogen peroxide decomposes rapidly, such that maintaining a precise concentration is difficult. We finally settled on a concentrated solution (5–6%) of sodium hypochlorite, otherwise known as Clorox. It killed all living things and destroyed any nucleic acids that are exposed. The problem then became how to kill the outer contaminating organisms, while protecting the interior organisms. We determined that full-strength (5.25%) Clorox for 10 seconds, followed by a series of rinses with sterile distilled water worked best. We also tested the method with DNA and RNA molecules that were not encapsulated in cells or virus coats, as well as intact viruses. The results were the same, Clorox treatments destroyed everything on the outside surface of the ice core sections (including *U. atrum*), but once rinsed from the surface, there were no detrimental effects on the organisms and nucleic acids within the ice core section.

Once we developed a dependable decontamination method, we could reach back further into the past with our ice core investigations. At first, only microscope and culture methods were used. In this process, growth media (i.e., food for the microbes) were placed into culture plates, tubes or flasks, and a small amount of the ice core meltwater was placed into each of the culture plates, which were then incubated at various temperatures. If the microbes grew, the colonies were visible after a few days to several months. Cells from the colonies then were examined using microscopes to identify the organism. More often than not nothing grew on a given plate. On average there usually was one colony per 20–200 plates. Some cells were removed for study by light microscopy, fluorescence microscopy, and/or electron microscopy. Some were examined using metabolic tests. These involved transfer of some cells onto various growth media to determine the growth requirements for that particular microbe. In general, some of the tests indicated genus and species of the organism. While inefficient, we isolated thousands of microbes.

Following isolation of the meltwater from the ice core interior, the process of characterization, including culturing and DNA sequencing can begin (Fig. 8.7). During these processes, not even a single cell, spore, virus particle, or stray molecule of DNA or RNA can be introduced into the samples. If bacteria, fungi, or other organisms are found, are they ancient bacteria or modern contaminants? If sterile procedures were employed throughout the research, then they probably are from the ice core section, but it is impossible to be 100% certain. There is always a chance that a stray cell was introduced as a contaminant. Therein lies the

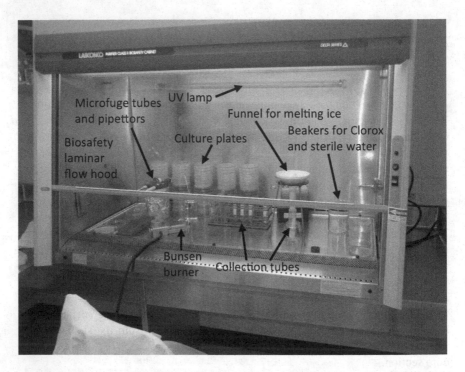

FIGURE 8.7 Arrangement of equipment for melting and culturing for ice core sections. All procedures are performed in a biosafety laminar flow hood. The hood is first cleaned with Clorox and ethanol, and exposed for 30 min to germicidal UV-irradiation. All of the equipment and media (previously autoclaved) are then placed into the hood. The ice core section is immersed in full-strength Clorox, and then rinsed three times in sterile reverse osmosis water. The core section is then allowed to melt at room temperature and fractions ("shells") are collected in tubes. Culture plates are also inoculated, each with 200 ul of the meltwater. Some of the meltwater is also collected for testing with PCR.

concern with all studies of ancient materials. Repetitions can be made to increase the chances that no contaminant is present, but there is still no way to be absolutely certain. Through stringent decontamination methods, we and others have gone to great lengths to ensure that the chances of isolating and characterizing ancient organisms rather than modern contaminants is maximized. We also perform isolations and sequencing in multiple replicates for each meltwater sample. Additional tests for quality control are routinely performed. By going through these elaborate protocols, we are assured that most, if not all, of our isolates and sequences are from ancient organisms in the ice.

MOST EFFECTIVE METHODS

A wide variety of methods have been developed to search for life in ice and permafrost (Fig. 8.8). These include microscopy, culturing, assaying for their

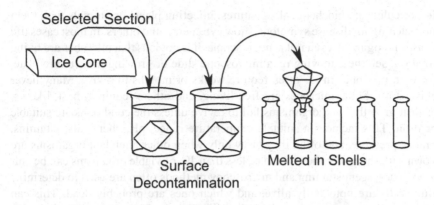

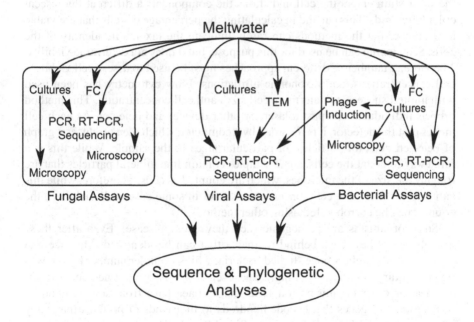

FIGURE 8.8 Some of the methods used to study life in ice. After the ice cores have been drilled from a glacier (or other source), the outer surfaces must be decontaminated in a research lab. In our lab, sodium hypochlorite (i.e., Clorox) is used. Then, the core section is melted in fractions (termed "shells"). The meltwater is then used for culture studies, PCR, RT-PCR, cDNA synthesis, sequencing (Sanger or next generation sequencing), metagenomics metatranscriptomics, and/or microscopy (light, fluorescence, scanning electron, and/or transmission electron microscopy). Other methods, including flow cytometry (FC), X-ray, CAT scans, and others can also be used.

biomolecular and biochemical signatures, infecting plants or animals with them and watching for disease symptoms, flow cytometry, and others. In most cases life is easily recognized, even if it is microscopic. It is obvious that microbes are living when you see them moving on a microscope slide or growing in a culture plate. However, microbes in ice often require weeks or months to revive. Many never revive at all. Some microbes are free living, and others require a host, like us, for their growth. Most organisms fail to revive unless the conditions are suitable for them. This includes a suitable temperature range, pH, chemicals, vitamins, osmotic pressure, and often they grow only when other attendant organisms are grown with them. For these reasons, less than 1% of viable organisms can be cultured. Fluorescent staining and microscopy methods often are used to determine which cells are potentially alive, and which ones are probably dead. This can be accomplished using a so-called "live-dead" stain. If the membrane is intact, one of the stains fluoresces (often green), whereas, if the membrane is broken, the other stain enters the cell and stains the components a different fluorescent color (often red). This can aid in calculating the percentage of cells that are viable in a sample, but the methods can rarely indicate the taxonomic identity of the cells. Sequencing can be used for this purpose, but it does not determine viability. Therefore, a number of different approaches must be used to assess cell viability, cell concentrations, and taxonomic affiliations. Flow cytometry has been used with limited success, to determine cell sizes and cell concentration. This method propels individual cells past a laser (or other device) and detector. As each cell moves past the detector, it is recorded by a computer, which then produces a graph of detected cells (or particles) of particular sizes in the sample. While this is a rapid way to count the cells, some samples contain fine mineral particles that are roughly the size of the microbes, which are counted as cells. Therefore, while this can help to determine cell concentrations (and in some cases cell viability), the counts must be corroborated using other methods.

Some organisms are pathogenic (i.e., they cause disease). Even after these microbes die, they leave behind remnants that can be identified. Most ice and permafrost researchers have studied bacteria, while a smaller number have investigated eukaryotes (usually, fungi), and even fewer have searched for viruses, archaea, or other organisms. Bacteria and archaea have from several hundred to thousands of genes that encode hundreds to thousands of proteins that carry out countless biochemical functions. So, although they are microscopic, they are nonetheless complex. Eukaryotic microbes are more complex than bacterial ones, possessing from several thousand to tens of thousands of genes within their nuclei, as well as dozens to hundreds in their organelles (mitochondria, present in most eukaryotes; and plastids, present in photosynthetic eukaryotes, such as higher plants and algae; and other more complex organelles, present in Alveolates, Excavata, Hacrobia, Rhizaria, and Stramenopiles; see Figs. 6.1 and 6.2). Viruses are less complex, having from just a few genes to more than 1,000 genes. However, because they are obligate parasites (they reside inside host cells), they generally utilize the products of many more genes that they use from the host cells. Although they are less complex than cellular life forms, they are difficult to study because they are small, only distantly related to one another, must grow in a host cell, and have a relatively small number of genes upon which to base identification. Less than a handful of virus species have been assayed from ice, and to date only a few researchers have even attempted to assay ice for viruses. Our labs have been investigating several viruses, including poxviruses, influenza A viruses, tobamoviruses (that infect plants), enteroviruses (including poliovirus), and a few others. With each viral group, one or more sets of specific molecular probes must be developed to find the viruses in the ice samples. Therefore, for each set of microbes there are unique sets of assays that must be performed to assure that as many species as possible are detected and identified.

SOURCES AND ADDITIONAL READINGS

Christner, B.C., J.C. Priscu, A.M. Achberger, C. Barbante, S.P. Carter, K. Christianson, A.B. Michaud, J.A. Mikucki, A.C. Mitchell, M.L. Skidmore, T.J. Vick-Majors, and the WISSARD Science Team. 2014. A microbial ecosystem beneath the West Antarctic ice sheet. Nature 512:310-313.

D'Elia, T., R. Veerapaneni, and S.O. Rogers. 2008. Isolation of microbes from Lake Vostok accretion ice. *Appl. Environ. Microbiol.* 74: 4962-4965.

D'Elia, T., R. Veerapaneni, V. Theraisnathan, and S.O. Rogers. 2009. Isolation of fungi from Lake Vostok accretion ice. *Mycologia* 101: 751-763.

Ma, L., C. Catranis, W.T. Starmer, and S.O. Rogers. 1999. Revival and characterization of fungi from ancient polar ice. *Mycologist* 13: 70-73.

Ma, L., S.O. Rogers, C. Catranis, and W.T. Starmer. 2000. Detection and characterization of ancient fungi entrapped in glacial ice. *Mycologia* 92: 286-295.

Petit, J.-R, J. Jouzel, D. Raynaud, N.I. Barkov, J.-M. Barnola, I. Basile, M. Bender, J. Chappellaz, M. Davis, G. Delaygue, M. Delmotte, V.M. Kotlyakov, M. Legrand, V.Y. Lipenkov, C. Lorius, L. Pépin, C. Ritz, E. Saltzman, and M. Stievnard. 1999. Climate and atmospheric history of the past 420,000 years from the Vostok ice core, Antarctica. *Nature* 399: 429-436.

Rogers, S.O. 1994. Phylogenetic and taxonomic information from herbarium and mummified DNA. In: R.P. Adams, J. Miller, E. Golenberg, and J.E. Adams (eds.) Conservation of Plant Genes II: Utilization of Ancient and Modern DNA. Missouri Botanical Gardens Press, St. Louis, pp. 47-67.

Rogers, S. O. 2017. Integrated Molecular Evolution, 2nd ed. Boca Raton, FL: Taylor & Francis Group.

Rogers, S.O., Y.M. Shtarkman, Z.A. Koçer, R. Edgar, R. Veerapaneni, and T. D'Elia. 2013. Ecology of subglacial Lake Vostok (Antarctica), based on metagenomic/metatranscriptomic analyses of accretion ice. *Biology* 2: 629-650.

Rogers, S.O., Theraisnathan, V., Ma, L.J., Zhao, Y., Zhang, G., Shin, S.-G., Castello, J.D., and Starmer, W.T. 2004. Comparisons of protocols to decontaminate environmental ice samples for biological and molecular examinations. *Appl. Environ. Microbiol.* 70: 2540-44.

Shtarkman Y.M., Z.A. Koçer, R. Edgar R, R.S. Veerapaneni, T. D'Elia, P.F. Morris, and S.O. Rogers. 2013. Subglacial Lake Vostok (Antarctica) accretion ice contains a diverse set of sequences from aquatic, marine and sediment-inhabiting Bacteria and Eukarya. PLoS ONE 8(7): e67221. doi:10.1371/journal.pone.0067221.

LINKS

https://icecores.org/
http://www.antarcticglaciers.org/glaciers-and-climate/ice-cores/ice-core-basics/
https://www.researchgate.net/figure/Map-of-Greenland-showing-the-position-of-ice-
    core-records-red-and-meteorological_fig1_286417720
https://www.sciencedaily.com/terms/antarctic_ice_sheet.htm
https://www.sciencedaily.com/terms/greenland_ice_sheet.htm

9 Isolating and Characterizing Microbes from Ice

"The answers you get . . . depend upon the questions you pose"

Margaret Atwood

HOW TO STUDY LIFE

We began to reexamine our definitions of life after we had been working with microbes and nucleic acids in ice for about a year. What had once seemed like simple questions: "What is life?" and "How do we find and identify microbes in ice?" became more difficult than we had thought previously. Some life could remain in a state of suspended animation and slowed metabolism for millennia, and longer. In this state the organisms may appear to be dead. We used several different ways to look for life in ice. First, we simply used light, fluorescence, and electron microscopes to try to find some recognizable shapes of fungi, bacteria, and/or viruses. Although we had some success at this, it was tedious and many long days resulted in little more than fatigued eyes. We tried concentrating the ice meltwater and inoculating plants and animals with the concentrates, with limited success. However, two methods were the most successful and efficient: culturing and nucleic acid sequencing, including metagenomic and metatranscriptomic approaches. These methods are used by several labs around the world to study microbes in ice. Approximately 90% of the research to examine microbes in ice currently involves these methods.

CULTURE METHODS (SEE FIGS. 8.7 AND 8.8)

Only a drop (0.1 to 0.2 ml, equivalent to 100 to 200 μl) of the ice meltwater is needed to inoculate each culture plate containing nutritive media with agar, or a tube full of a liquid nutrient broth. These are incubated for a few weeks to many months at temperatures usually between 4° and 22°C. This method is tedious and technically challenging, due to concerns with strict avoidance of contamination, and because growing organisms are found in only one plate or tube in 20 to 200. However, if the organisms grow, they multiply into millions or billions of copies, and they are then much easier to study, including yielding a great amount of sequence data. In our laboratories, we have isolated thousands of bacteria and fungi from ancient ice cores this way. Worldwide, there are many times this number that have been isolated in other research laboratories. However, the majority of organisms in the ice are nonculturable with present techniques, and therefore we are only able to identify a small select group of organisms using those methods. It is estimated that currently scientists are able to culture only 0.1% to 1.0% of the viable organisms that are present in a particular sample. Therefore, a gross underestimate of the total number of viable cells is calculated by this method. Another limitation is that we rarely know exactly what organisms to expect. This is important, because the growth media, incubation temperature, and other conditions cannot be reliably predicted and employed. Also, personnel may be unknowingly exposed to pathogens in the ice. Thus, often it is educated guesswork to choose the specific growth conditions to use and which precautions to employ. Nonetheless, culturing has provided a large stock of microbes for more detailed studies, and no known diseases or deaths have yet been reported from these studies. We hope that these trends will continue.

NUCLEIC ACID METHODS

During the past several decades, molecular biology methods have become powerful tools in searching for life in ancient matrices, including ice and permafrost. The two methods used most often are PCR (polymerase chain reaction) and sequencing. PCR (Fig. 9.1) can be used to amplify one or a few molecules into millions of copies in a few hours. The copies then can be used to determine the sequences of the original molecules, which then can be compared to known sequences in national and international databases to determine the identity of the species in the samples. Individual or sets of genes can be scanned, or genomic sequencing can be performed on especially important organisms. Each of the molecular methods can be so sensitive that single molecules can be detected and amplified. They can be accurate in identification of genera, species, strains, and varieties of organisms. For nucleic acid sequencing, only 1 to 10 μl of the ice meltwater is needed for each assay. For PCR amplification of DNA from bacteria, archaea, or fungi, short pieces of synthesized DNA (termed "primers," because they prime, or initiate, the reaction) are utilized that are composed of sequences that will adhere (hybridize) to specific locations on the template DNA in the samples (if the specific DNA is present). Alternatively, random primers can be used to amplify all of the sequences in a sample, in a process known as metagenomics analysis. Next, a DNA polymerase (that makes new DNA based on the sequence of the template DNA) is added. Other components also are added and the reaction begins. The synthesized pieces are designed so that a specific region of specific length will be copied many times over. These copies can then be used to determine the DNA sequence of the original template molecule. This is a powerful technique, because it is very sensitive and specific. Identification to the species level can be achieved within a day or two. However, with increasing sensitivity comes a concurrent sensitivity to contamination. PCR does not discriminate between the target nucleic

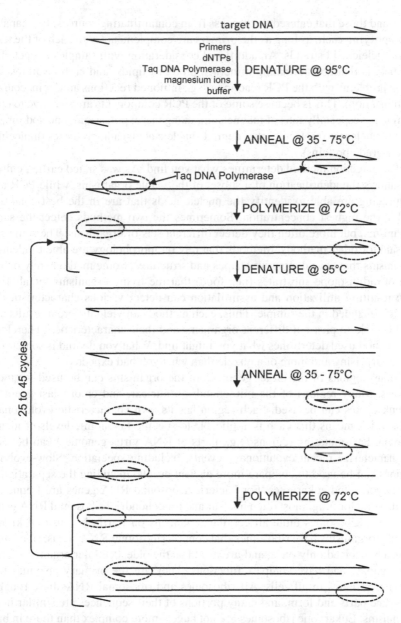

FIGURE 9.1 Polymerase chain reaction (PCR). DNA primers, dNTPs (deoxyribonucle-otides dATP, dCTP, dGTP, and dTTP), buffer (usually at pH 8.3), Taq DNA polymerase, and magnesium ions (Mg^{2+}) are added to DNA, in order to amplify a target sequence in the DNA. The mixture is heated to 95°C to denature the DNA, followed by lowering the temperature so that the primers will anneal to the complementary sequences. The temperature is then raised to 72°C, which is the optimal temperature for Taq DNA polymerase. The temperature shifts are then repeated from 25 to 45 times, which repeatedly amplifies the target sequence.

acids and those that entered the process from contaminating sources. It is capable of amplifying each, as long as the primers are complementary to each of the template nucleic acid strands. An additional consideration with samples collected in the field is that contaminating minerals, proteins, lipids, and carbohydrates can cause problems with the PCR reactions (e.g., inhibited reactions and/or inaccurate amplification). This is because some of the PCR components are sensitive to contaminants, especially ions of magnesium, manganese, potassium, and sodium, as well as nucleases (molecules that degrade nucleic acids) and proteases (molecules that degrade proteins).

The detection method determines what you find. As was stated earlier, culturing allows the identification of a subset of the viable organisms, while PCR and sequencing are able to identify the nucleic acids that are in the best condition and at the highest concentration. Sometimes the two methods detect the same organisms, but more often they detect different sets of organisms. The same can be said for other methods. Methods that rely on morphology are able to identify organisms that have distinctive shapes and structures. Some methods rely on the use of radioisotope substrates (i.e., food) that the living organisms metabolize. The resulting utilization and assimilation can detect various characteristics of the life included in the sample. Thus, each method can yield different results and will be able to pick out different organisms and their characteristics. Therefore, the method used determines what you might find. What you do find is sometimes more surprising and more of a puzzle than what you had expected.

Many regions of the target genomes of the organisms can be used for these studies. Many regions of the mitochondrial, nuclear, and chloroplast genomes of eukaryotes can be used. Each region has its own characteristic evolutionary rate, which means that each is applicable to specific taxonomic levels of identification. Fast-mutating regions (e.g., parts of RNA virus genomes) can be used to characterize recent evolutionary events, including speciation. Slow-evolving regions can be used to compare more ancient events, including the separation of prokaryotes and eukaryotes. For bacteria, ribosomal RNA genes are frequently used, while for fungi most often nuclear and mitochondrial ribosomal RNA genes are used. Ribosomes unite all cellular organisms on Earth, because all known cellular organisms have ribosomes that contain ribosomal RNA. Thus, ribosomes probably evolved only once, and are as old as the oldest cellular organism. They are complex and have complex functions. They are the primary enzymes that produce proteins in all cells. All ribosomes and ribosomal RNAs have roughly the same parts and form, and many portions of their sequences are similar in all organisms. Eukaryotic ribosomes are not hugely more complex than those in bacteria, and the mitochondrial and chloroplast versions are very much like their bacterial counterparts, although they are somewhat smaller. There are many different regions of the ribosomal genes that each evolves at a different rate. They have been used extensively in molecular evolutionary studies to answer questions of genealogy for the major groups of organisms on Earth. In fact, the small subunit genes have been used to examine the entire tree of life, from primitive bacteria to the most complex eukaryotes (the tree in Fig. 6.1 is based on sequences of this gene from a wide range of organisms). On the other hand, the gene spacer regions have been used to examine relationships at the species level. It is possible that ribosomes and ribosomal DNA occur in extraterrestrial locations. If organisms with ribosomes, or rDNA, are found in extraterrestrial locations, it would indicate a common source for Earth and extraterrestrial organisms. If ribosomes are ever found on another planet or moon, did they originate on Earth or elsewhere?

When we first started isolating and sequencing these organisms it was more-or-less a blind effort, not knowing what we would find. Of course, we could make educated guesses about the general types of organisms that we might find, but it was a shotgun approach. Once we had a large number of cultures and sequences, we could finally start looking for trends in the types of microbes represented in the ice. The first conclusion was that there were many organisms, and the diversity was higher than expected. But, the vast majority were bacteria. The second was that many of the organisms found in the ice are floating around in the air that we breathe every day. The startling aspect was that some were almost identical to those that were entrapped in 500,000-year-old ice. On the other hand, some microbes found in ice were unknown to science. Even after two decades of research, some still remain as unknowns. No one knows at this point whether these simply have never been studied, or whether they represent extinct species.

METAGENOMICS, METATRANSCRIPTOMICS, AND GENOMICS

During the past two decades, we changed our strategy, primarily because newer methods led us on a different path, and provided much more data. Metagenomics and metatranscriptomics have been used to determine the entire genetic contents of ice samples (Fig 9.2). In some cases, a combination of the two has been employed. Metagenomics is a method whereby all of the DNA in a sample is amplified using either specific or random (i.e., non-specific) primers to initiate the polymerization reactions (Fig. 9.2). The resulting amplified fragments can then be sequenced using one of the next generation sequencing (NGS) methods. These usually produce hundreds of thousands to billions of sequences (called "reads"). Using bioinformatic methods, the sequences can be characterized, grouped, and subjected to searches using personal, national, or international sequence databases. This can be used to identify species affiliations, biochemical pathways, physiology, ecology, etc. Metatranscriptomics detects all of the RNA in a sample, which represents the genes that are being expressed by the organisms in the ice. In this method, the RNA is subjected to cDNA synthesis, which is a process to make DNA copies of the RNA. The resulting cDNAs are then amplified using PCR, sequenced, and analyzed, as above for metagenomics. As with the other methods that use amplification of the nucleic acids, contaminants also can be amplified in the process, so careful decontamination of the outer surfaces is crucial to obtaining reliable results. While culturing methods allowed us to identify hundreds of living microbes in the ice samples, the metagenomic and metatranscriptomic results provided information about the tens of thousands of microbes in the ice

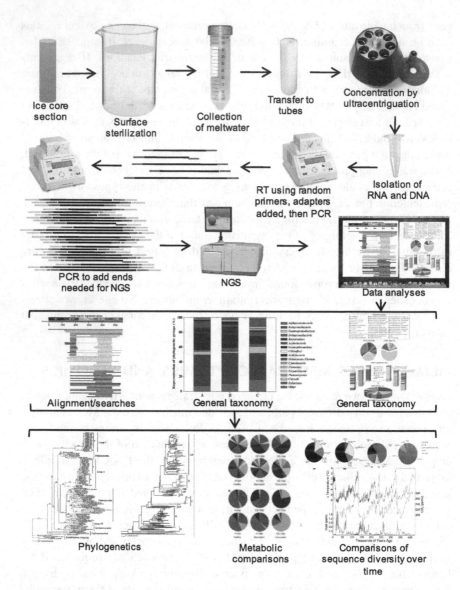

FIGURE 9.2   General metagenomics method. The ice meltwater is first subjected to ultracentrifugation to concentrate the nucleic acids and organisms in order to have sufficient amounts of nucleic acids. The RNA is used to produce cDNA copies using reverse transcriptase (RT) and random primers. Known sequence pieces are added to the ends so that PCR can be used to amplify all of the fragments. Next, additional specific sequences are added to the ends of all of the fragments, such that next generation sequencing (NGS) can be performed. Once the sequences are determined, analyses are performed to determine taxonomy (e.g., species, genera, families, phyla, kingdoms), phylogenetics to determine fine structure relationships, metabolic pathways to determine what the organisms are doing in the ice, and other analyses.

samples, and also gave us a great deal of information of the functions they were performing in their environments and habitats.

The metagenomic data give slightly different, although overlapping, information than does the metatranscriptomic data. The metagenomic sequences will yield information regarding all of the species in the sample, as well as some of the metabolic capabilities of those organisms. The metatranscriptomic sequences will indicate primarily the species that are living and transcribing rRNA, and will indicate the metabolic processes that were active in the sample. If random primers are used in metatranscriptomics, approximately 80% of the sequences will be from various sections of the rRNAs. This is because most of the RNA in cells is rRNA, approximately 80% (by mass) for most cells. The other RNAs represent RNAs for various cellular processes that are being expressed at the time the sample was assayed. The sequence profile here provides taxonomic information that is from the rRNA sequences, as well as gene expression information, which can determine what the cells were doing metabolically while entombed in the ice, or just prior to being frozen. When metagenomics and metatranscriptomics are employed concurrently on a meltwater sample, a great deal of information is gleaned. Species affiliations can determine the profile of taxa in the sample, and their potential biochemical functions. For example, nitrogen-fixing bacteria, nitrifying bacteria, carbon fixing bacteria, and other metabolic capabilities, can be determined in many cases. This can be confirmed by searching for sentinel gene sequences in the sample to build a picture of the biochemical pathways, cycles, and processes within an ice sample.

Genomics can determine an entire genome, or a portion thereof, of a particular organism. In some cases, a genome can be determined from a single cell. However, because of the technical difficulties and the costs, this has not yet been attempted for an ice sample. If the identity of a particular cell is unknown, it is difficult to subject it to genomic sequencing. A clearer path to genomic sequencing is to culture the organism first. In this way, the identity is at least partially known, and a large quantity of DNA can be obtained for genomic sequencing. But, as previously mentioned, only about 1% of the viable cells in a sample can be successfully cultured, so only a small proportion of the genomes in an ice sample can be effectively determined. Viruses have relatively small genomes, so it might be possible to sequence a number of these. However, viruses exhibit great genetic variability, so only specific genes and specific viruses are amenable to sequencing. Even then, complete genomes may be difficult to achieve because of the variability. An additional disadvantage to studying viruses in ice is that when they are outside of a cell, any damage to their nucleic acids is unrepaired. However, if they are still in a host cell, either as a virus or as a prophage (integrated into the host chromosomes), repair is possible, because some organisms are able to metabolize (at a low rate) while entombed in ice. If the host cells can be cultured, it is possible to induce a virus to switch from a lysogenic phase (as a provirus) to a lytic or active phase by using chemicals, UV-irradiation, or temperature shocks. This is a process whereby the virus chromosome excises from the host chromosome, and then the genes are expressed to form new virus particles to be released from the host cell. In this way, a large amount of viral RNA or DNA can be isolated and sequenced. We induced the lytic switch in one such virus that was identified as a PBSY virus of *Bacillus subtilis* isolated from Greenland ice cores. However, the resultant virus particles were defective in most cases, indicating some damage to the virus genome had occurred while it was a lysogen in the entombed bacterium.

OTHER METHODS

Additional methods have been used or proposed to assay for microbes in the environment, including ancient ice. Indicators of life, life processes, and signs of fossil life range from fairly simple and straightforward to technically challenging, ambitious, or ambiguous. For centuries, scientists have described and identified microbes using microscopes. One can determine whether or not microbes are present, as well as discerning the shape, quantity, and ornamentation (surface features) of the cells present in the sample using microscopy. However, often this only gives a rough idea of the type of organism, and some mineral formations can resemble microbes. Some chemical stains can increase the precision of identification. So-called "vital" stains can be used to determine whether or not a particular cell is alive or dead. Living cells will bind the stain, whereas dead ones will not. The only problem is that most of these stains eventually kill the cells. A major problem with microscopy methods is that some ice cores contain so few organisms that it is extremely difficult to find the microbes, and experts are needed to identify them. Often, they are attached to pieces of sand and soil, and therefore cannot be easily located and examined. Some ice core researchers concentrate the samples by filtering the meltwater, while others use centrifuges and then examine the pelleted material. While these methods increase the likelihood of finding the microbes, they can also kill some of them.

Another method to detect living organisms is by autofluorescence. This method is used to detect the concentrations of specific chemicals or viable cells, but only generally because it cannot discern individual taxonomic groups. Several compounds in living cells, primarily those containing molecular ring structures (e.g., proteins containing the amino acids tryptophan, tyrosine, and phenylalanine; and compounds containing nucleotides and nucleosides, such as purines and pyrimidines) fluoresce at specific wavelengths (often colors in the visible range) when exposed to higher energy wavelengths of light. A laser or ultraviolet light source is shone on a sample and the amount of fluorescence of the characteristic wavelengths is measured. For some, the fluorescence occurs only if the cells are alive. The concentrations of cells must be fairly high with some fluorescence methods, in order to detect the presence of the cells, although some improvements have yielded increases in sensitivity. Some researchers have designed sophisticated robots that can be lowered into the ice core boreholes that can record the amount of fluorescence from living microbes as they are lowered into the holes.

Fluorescent stains have also been developed that can determine whether a cell is alive (by fluorescing one color because the cell membrane is intact, which excludes the dye from the interior of the cell) or dead (by fluorescing a different color because the membrane is broken, which allows the second dye into the cell). These methods cannot be used to identify a microbe to species, but are useful for determining viability. Other biological compounds can be measured using additional methods (e.g., the measurement of total organic carbon, TOC; dissolved organic carbon, DOC; chlorophyll a; NADH; DNA; or other biological indicator molecules). These methods provide estimates of the concentrations of living plus dead organisms in the samples, because the compounds are preserved in the ice long after the organisms die. Many have suggested assaying for methane production, which often indicates biological activity.

Scanning and transmission electron microscopy have been used to observe the outer surfaces of cells, and in some cases the cells can be identified to genus and species this way. Another way to look at cell surfaces is by atomic force microscopy. Atomic forces cause minute deflections in a magnetic field that are interpreted by computer to form an image of the surface atoms and molecules on the cell surface. One other method comes by way of the medical profession. When a doctor wants to view parts of the inside of your body in detail, there are several methods available, such as MRI (magnetic resonance imaging), CAT (computer automated tomography) scans, and X-rays. These same methods have been used to examine samples of soil, rock, and ice. Some are of sufficient clarity and resolution, that individual bacteria can be visualized. However, in order to achieve higher resolution and deeper penetration of the samples, high energies are needed. As the energies are increased, the likelihood of killing the included organisms and destroying biological molecules also increases. Such methods are being investigated for possible inclusion on future space probes to search for life on Mars and elsewhere.

While the methods used to isolate, or otherwise identify microbes, must be sensitive and accurate, sensitivities can be problematic. If a method can detect a single cell in a liter of water, then there can be no contaminating organisms in that water to interfere with the assay. Therefore, if sensitive methods are used, then thorough methods to eliminate contamination also must be employed. Ignoring this issue leads to failure. First, the results cannot be trusted. This means all of the results, including any results that actually may be accurate, have to be thrown out. Second, the scientific community will refuse to accept the results in present and future work if any of the results are found to be contaminants. Third, the entire community of researchers loses credibility with every study that is found to be inaccurate or erroneous. The stakes are high, and thus researchers take this issue very seriously. The current methods reflect decades of research that have led to a refinement and improvement of the various methods. With increasing sensitivity has come improved decontamination methods, more attention to detail, and results that can stand up to criticism, scrutiny, and challenge. The field of examination of ancient materials can now be focused on testing of reliability, accuracy, and repeatability. If the results can be replicated in one or more laboratories, then they can be trusted. Although this is true for all scientific studies, it is crucial to studies involving ancient specimens. The future studies being planned are also paying more attention to avoidance of contamination of the sample sites.

SOURCES AND ADDITIONAL READINGS

Castello, J.D., and S.O. Rogers. 2005. *Life in Ancient Ice*. Princeton, NJ: Princeton University Press.

D'Elia, T., R. Veerapaneni, and S.O. Rogers. 2008. Isolation of microbes from Lake Vostok accretion ice. *Appl. Environ. Microbiol.* 74: 4962-4965.

D'Elia, T, R. Veerapaneni, V. Theraisnathan, and S.O. Rogers. 2009. Isolation of fungi from Lake Vostok accretion ice. *Mycologia* 101: 751-763.

Ma, L., C. Catranis, W.T. Starmer, and S.O. Rogers. 1999. Revival and characterization of fungi from ancient polar ice. *Mycologist,* 13:70-73.

Ma, L., S.O. Rogers, C. Catranis and W.T. Starmer. 2000. Detection and characterization of ancient fungi entrapped in glacial ice. *Mycologia* 92: 286-295.

Rogers, S.O. 1994. Phylogenetic and taxonomic information from herbarium and mummified DNA. In: R.P. Adams, J. Miller, E. Golenberg, and J.E. Adams (eds.) *Conservation of Plant Genes II: Utilization of Ancient and Modern DNA*. Missouri Botanical Gardens Press, St. Louis, pp. 47-67.

Rogers, S.O. 2017. *Integrated Molecular Evolution*, 2nd ed. Boca Raton, FL: Taylor & Francis Group.

Rogers, S.O., Y.M. Shtarkman, Z.A. Koçer, R. Edgar, R. Veerapaneni, and T. D'Elia. 2013. Ecology of subglacial Lake Vostok (Antarctica), based on metagenomic/metatranscriptomic analyses of accretion ice. *Biology* 2: 629-650.

Rogers, S.O., Theraisnathan, V., Ma, L.J., Zhao, Y., Zhang, G., Shin, S.-G., Castello, J.D., and Starmer, W.T. 2004. Comparisons of protocols to decontaminate environmental ice samples for biological and molecular examinations. *Appl. Environ. Microbiol.* 70: 2540-44.

Shtarkman Y.M., Z.A. Koçer, R. Edgar R, R.S. Veerapaneni, T. D'Elia, P.F. Morris, and S.O. Rogers. 2013. Subglacial Lake Vostok (Antarctica) accretion ice contains a diverse set of sequences from aquatic, marine and sediment-inhabiting Bacteria and Eukarya. PLoS ONE 8(7): e67221. doi:10.1371/journal.pone.0067221.

10 A Brief History of Research on Life in Ice

"I was a victim of a series of accidents, as are we all"

Kurt Vonnegut Jr.

LONGEVITY

In past centuries (and even today) extracts of ancient organisms were thought to confer longevity to those who ate them. Others (like us) simply have an innate curiosity about how and why organisms are able to survive for long periods of time. There are obvious advantages to being successfully frozen and thawed. Some people have had themselves frozen, with the hope that decades from now they can be brought back to life and cured of any diseases that they had, including old age. Good luck to them. Unfortunately, currently, humans cannot be frozen and resuscitated after thawing. However, some animals (e.g., some insects, amphibians, and crustaceans) can survive freezing and thawing, as can some plants. Lotus seeds remain viable for at least several centuries, while estimates on the longevity of bacterial spores ranges from 1000 to 250,000,000 years. Water bears (tardigrades) have survived freezing to nearly absolute zero (-273°C, -459°F). Some nematodes that had been frozen in permafrost for more than 40,000 years, when thawed, began moving and eating. Since the mid 1990's, thousands of organisms (Figs. 10.1, 10.2, 10.3, and 10.4) and millions of DNA sequences (Figs. 10.5 and 10.6) have been detected in ice cores from Greenland and Antarctica that are from hundreds to millions of years old, and from glaciers in temperate and equatorial zones. Current estimates of the total number of viable microbes melting from the ice annually range from 10^{18} to over 10^{22}, based on studies of ice cores and the demonstration that more than 500 billion tons of ice is melting annually worldwide. The information so far gleaned from the studies of microbes in ice has led to some important findings. First, wind-transported microbes and tissues, including pollen grains, from virtually anywhere on Earth have been detected in glaciers worldwide. Second, ice is an excellent (possibly the best) preserver of microbial life and biological macromolecules. Third, the organisms that can survive freezing and thawing in glaciers have the potential to emerge from glaciers millennia (or longer) after they were deposited, whereupon they may interact with extant populations, or infect extant hosts, including humans. Although

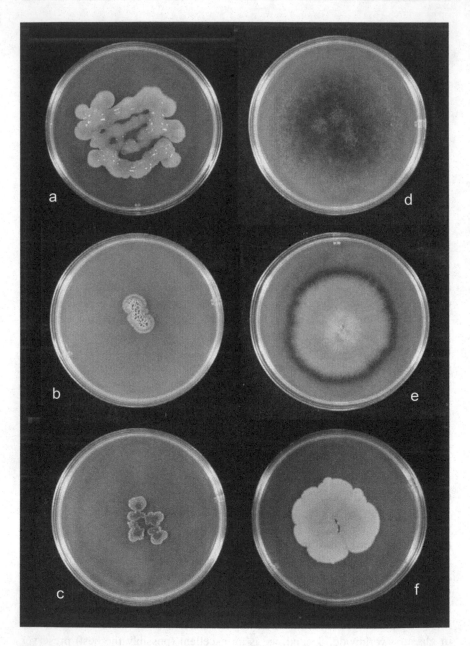

FIGURE 10.1 Examples of bacteria and fungi revived from 500 to 7,000-year-old Greenland ice cores. **a.** *Rhodococcus* sp. (bacterium). **b.** *Actinobacterium* sp. **c.** *Rhodotorula minuta* (fungus). **d.** *Ulocladium atrum* (fungus). **e.** *Cladosporium cladosporioides* (fungus). **f.** unidentified fungus.

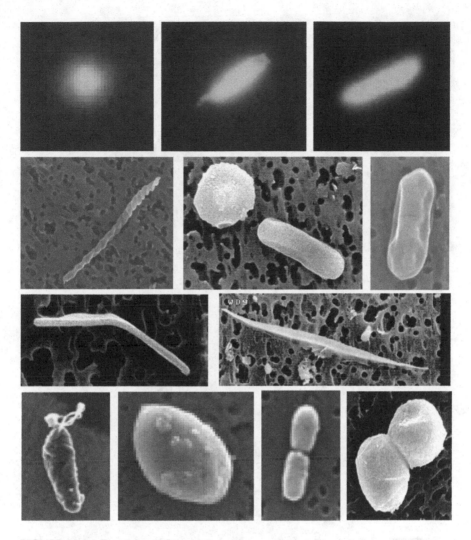

FIGURE 10.2 Examples of fluorescence micrographs (top three) and scanning electron micrographs (lower micrographs) of bacterial cells found in some of the Lake Vostok accretion ice core sections. Spherical, rod-shaped, and spiral cells are present, as well as a few other shapes.

speculative at this time, some disease outbreaks may support this speculation. Long-term infectivity for some viruses has been experimentally demonstrated. Fourth, the methods developed can be used to search for life in other parts of the Solar System. Fifth, some of the microbes found in ice may live there as their normal habitats, thus expanding our view of cold biospheres, and the ability of life to thrive in almost every conceivable habitat.

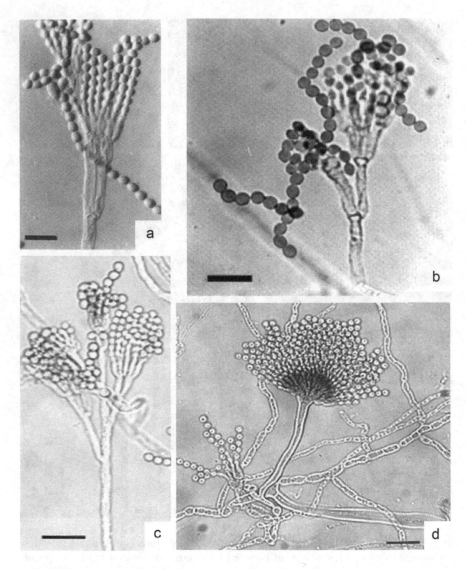

FIGURE 10.3 Micrographs of fungal isolates (*Penicillium* and *Aspergillus*) from Greenland ice core sections. **a.** *Penicillium* sp. from a 500-year-old core section. **b.** *Penicillium* sp. from a 600-year-old core section. **c.** *Penicillium* sp. from an 8,500-year-old core section. **d.** *Aspergillus* sp. from a 2,000-year-old core section.

A BRIEF HISTORY

Imagine a meeting between a group of explorers, sea captains, admirals, geographers, mycologists, climatologists, glaciologists, drilling crew members, chemists, biochemists, naturalists, biologists, and a friar. And, they are from the US, UK, Russia, Denmark, Norway, France, Italy, Austria, Czechoslovakia, Canada,

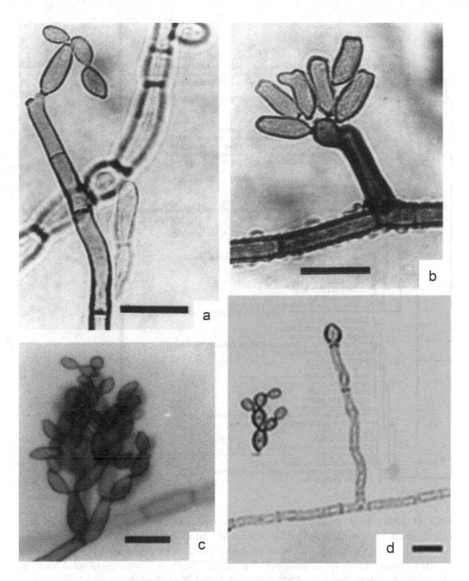

FIGURE 10.4 Micrographs of fungal isolates (*Cladosporium*) from Greenland ice core sections. **a.** *C. cladosporioides* from a 700-year-old core section. **b.** *C. herbarum* from a 110,000-year-old core section. **c.** *C. cladosporioides* from a 4,500-year-old core section. **d.** *C. cladosporioides* from a 3,500-year-old core section.

Denmark, Australia, Chile, and other countries. But, now imagine that all of these people were in different places at different times in history, but their ideas and discoveries were primarily in the form of publications. The explorers were people like Erik the Red (Viking) who was among the first Europeans to settle in Greenland (985 CE); Admirals Fabian Gottlieb von Bellinghausen and Mikhail

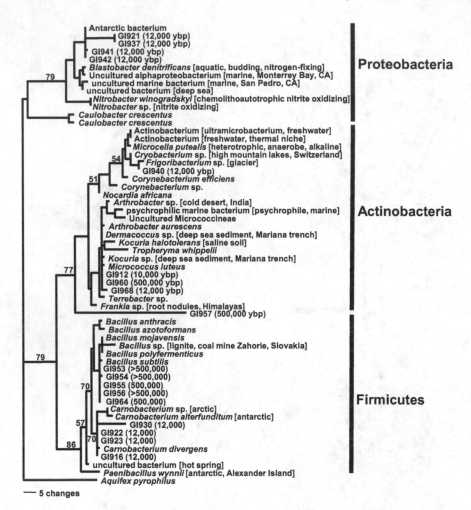

FIGURE 10.5 Evolutionary tree based on sequences from bacterial cultures from glacial ice cores, compared to modern sequences of related species. Sequences from cultures
are indicated by "GI" (glacial isolate) numbers. Approximate age of each ice core section
is noted in parentheses (ybp, years before present). The most commonly isolated bacteria
are within the phyla, Actinobacteria, Firmicutes, and Proteobacteria.

Lazarev (Russia), who first set foot in Antarctica (1820). Others followed with
puzzle pieces, such as, geologist Charles Lyell (UK), who determined that layers
of sedimentary rock represent sequential ages in the past (1830); Robert FitzRoy,
who was Captain of the Beagle during the voyage with Charles Darwin (1831–
1835); naturalist Charles Darwin (UK), who described the emergence of species by natural selection (1859); friar Gregor Mendel (Czech), who discovered
the characteristics of inheritance (1865); biologists Tschermak, deVries, Correns,

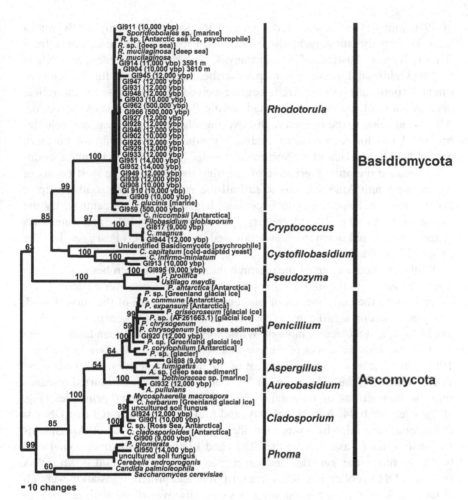

GI911 (10,000 ybp)
Sporidiobolales sp. [marine]
R. sp. [Antarctic sea ice, psychrophile]
R. sp. [deep sea)]
R. mucilaginosa [deep sea]
R. mucilaginosa
GI914 (11,000 ybp) 3591 m
GI904 (10,000 ybp) 3610 m
GI945 (12,000 ybp)
GI947 (12,000 ybp)
GI931 (12,000 ybp)
GI948 (12,000 ybp)
GI903 (10,000 ybp)
GI962 (500,000 ybp)
GI966 (500,000 ybp)
GI927 (12,000 ybp)
GI928 (12,000 ybp)
GI946 (12,000 ybp)
GI902 (10,000 ybp)
GI926 (12,000 ybp)
GI929 (12,000 ybp)
GI933 (12,000 ybp)
GI951 (14,000 ybp)
GI952 (14,000 ybp)
GI949 (12,000 ybp)
GI939 (12,000 ybp)
GI908 (10,000 ybp)
GI910 (10,000 ybp)
GI909 (10,000 ybp)
R. glucinis [marine]
GI959 (500,000 ybp)
C. niccombsii [Antarctica]
Filobasidium globisporum
GI817 (9,000 ybp)
C. magnus
GI944 (12,000 ybp)
Unidentified Basidiomycete [psychrophile]
C. capitatum [cold-adapted yeast]
C. infirmo-miniatum
GI913 (10,000 ybp)
GI895 (9,000 ybp)
P. prolifica
Ustilago maydis
P. antarctica [Antarctica]
P. sp. [Greenland glacial ice]
P. commune [Antarctica]
P. expansum [Antarctica]
P. griseoroseum [glacial ice]
P. sp. (AF261663.1) [glacial ice]
P. chrysogenum
P. chrysogenum [deep sea sediment]
GI920 (12,000 ybp)
P. sp. [Greenland glacial ice]
P. corylophilum [Antarctica]
P. sp. [glacier]
GI898 (9,000 ybp)
A. fumigatus
A. sp. [deep sea sediment]
Dothioraceae sp. [marine]
GI932 (12,000 ybp)
A. pullulans
Mycosphaerella macrospora
C. herbarum [Greenland glacial ice]
uncultured soil fungus
GI924 (12,000 ybp)
GI901 (10,000 ybp)
C. sp. [Ross Sea, Antarctica]
C. cladosporioides [Antarctica]
GI900 (9,000 ybp)
P. glomerata
GI950 (14,000 ybp)
uncultured soil fungus
Cerebella androprogonis
Candida palmioleophila
Saccharomyces cerevisiae

Rhodotorula
Basidiomycota
Cryptococcus
Cystofilobasidium
Pseudozyma
Penicillium
Aspergillus
Aureobasidium
Ascomycota
Cladosporium
Phoma

100
99
85
6?
97
100
100
100
100
100
99
59
100
64
54
100
100
89
100
85
99
60

= 10 changes

FIGURE 10.6 Evolutionary tree based on sequences from fungal cultures from glacial ice cores, compared to modern sequences of related species. Sequences from isolates are indicated by "GI" (glacial isolate) numbers. Approximate age of each ice core section is noted in parentheses (ybp, years before present).

and Spillman, who confirmed Mendel's earlier results (1900); and many others. It literally took an army of people from different countries, at different times, in different fields and occupations, to develop the studies of microbes in ice to the point they have reached today. We are merely privates in this army.

Although brief incursions into the Southern Ocean occurred prior to the 1800's, parts of Antarctica were not discovered until the early 19th century, and major exploration of the continent were only undertaken in the 1890's and early 20th century. Roald Amundson reached the geographic South Pole in 1911, and Robert Falcon Scott reached the same point only a few weeks later. During the subsequent three decades, explorers and researchers from Britain, Russia, Australia, US, Germany, Chile, Norway, Sweden, and others mounted additional missions to map, describe, and understand this frozen continent. Expeditions into the Arctic began much earlier than those to Antarctica, because most of the nations that had sailing fleets were closer to the Arctic. Also, in addition to the interest in discovering, claiming, utilizing, and colonizing new lands for each country, finding a northern route around the northern part of North America (the "Northwest Passage") was desired by many countries, because the only alternative to reaching the orient or the west coasts of North and South America, was to sail all the way around the southern tip of South America. Expeditions to Greenland began by the 13th century. By the end of the 19th and early 20th centuries, all of the islands, bays, and waterways in the Arctic were known in great detail, and in 1909, Robert Peary reached the geographical North Pole.

While the modern part of the research into the field of microbes in ice began a few decades ago, the first hints appeared in the literature over a century ago. In addition to the early studies of microbes in glaciers, all of the current work was predicated on significant discoveries in microbiology, genetics, evolutionary biology, and molecular biology. While some of the initial foundations of the work began with the works of Gregor Mendel and Charles Darwin in the mid-19th century, it was the establishment of the field of genetics in the 1910's and 1920's, following the rediscovery of Mendel's work, that jump started research into the interactions of inheritance factors and evolutionary principles (Fig. 10.7). Then, in 1944, Avery, McCarty, and McLeod demonstrated that DNA is the genetic material, which was confirmed in 1952 in another set of experiments by Hershey and Chase. During the 1950's and 1960's all of the basic foundation blocks of molecular genetics and genetic engineering were laid down. DNA structure, DNA replication, RNA transcription, and protein translation were all elucidated. Enzymes to cut and join DNA were discovered and analyzed. Pieces of DNA that could move around inside the cell (transposons and some introns) and move from cell to cell (viruses, phages, and plasmids) were discovered. In the 1970's more groundwork was laid, including the development of cloning methods, DNA sequencing, and in the 1980's, polymerase chain reaction (PCR) techniques of all sorts, and then by the 2000's, next generation sequencing (NGS) methods.

Since the 1950's many glaciologists, geologists, and climatologists laid the foundations for biological studies (Fig. 10.7). They mapped, drilled, measured, and prodded glaciers, ice domes, and ice fields throughout the world, including many investigations of glaciers, ice sheets, ice domes, ice shelves, and sea ice in and around Greenland and Antarctica. They measured the accumulation, melting, fracturing, temperature, thickness, speed, and consistency of the ice. They measured gases, minerals, sediments, and other inclusions. Through these studies they

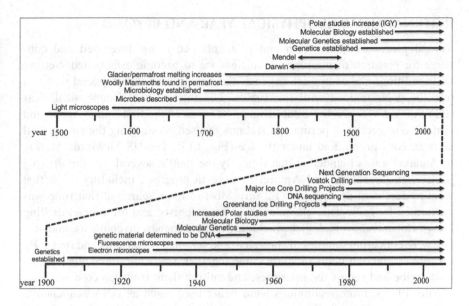

FIGURE 10.7 Timelines of some of the important developments that led to research into microbes in ice. The top timeline outlines the important discoveries and developments during the last 500 years that contributed to the study of organisms in ice, leading to the field of ice microbiology. This includes advancements in microscopy, genetics, and molecular biology. The lower timeline shows more details of the developments and discoveries over the past century. In particular, the coincidence of the increases in polar research, microbiology, and molecular biological methods greatly contributed to the current state of research.

determined the age of the ice at various depths, and the temperature of the Earth's atmosphere hundreds of thousands of years into the past. This was done by measuring isotopes of several elements and other markers in the ice that are correlated to age and temperature at the time of snow and ice deposition. The researchers made precise determinations of when there were ice ages and warming (interglacial) periods, as well as the temperature and gas fluctuations within each of the periods. Volcanic and other geological events were plotted. They found rocks in glaciers that came from the Moon and Mars. They plotted the amount of heating of the Earth compared to the levels of several atmospheric gases, and concluded that the current CO_2 (carbon dioxide), CH_4 (methane), and N_2O (nitrous oxide) levels are higher than they have been in at least the last 800,000 years. Given the current sustained rise in CO_2 and CH_4 (both powerful greenhouse gases) levels, global climate change, including global warming of the oceans, land, and atmosphere, will increase (perhaps uncontrollably) during our lifetimes. The intense study of glaciers provided a framework for the ensuing biological studies, without which they would never have been possible.

INTERNATIONAL GEOPHYSICAL YEAR AND BEYOND

Several pieces of information and protocols had to be developed and converge for research of microbes in ancient ice to become established. Setting up expeditions and camps are expensive endeavors, and this impeded research into the polar regions until the 1950's. The International Geophysical Year (IGY) of 1957/58 saw the establishment of some of the first major Arctic and Antarctic projects and permanent stations focused on studying the ice, as well as organisms in, on, and under the ice (Fig. 10.7). The US McMurdo Station in Antarctica was founded at that time. By the 1970's, several ice core drilling projects in Greenland and Antarctica were in progress, including some that went to depths of up to 2 km (1.2 mi). Most of the research at that time was performed by geologists, glaciologists, climatologists, and (of course) drilling crews. Concurrently, biologists, including bacteriologists, mycologists, and others, were collecting samples of snow, ice, and scrapings from soils and rocks. By the 1980's, the two fields of study came together, when microbiologists began to describe and report the microbes, and culture them from ice core sections. By the 1990's, more techniques were being used, such as DNA sequencing, fluorescence microscopy using stains to indicate whether the microbes were alive or dead, and measurements of ions and organic compounds in the ice. By the 2000's, next generation sequencing (NGS) methods, including metagenomic and metatranscriptomic analyses, were being employed. All led to similar conclusions: glacial ice contains a large number and diversity of microbes, and many of them are alive.

As mentioned above, the history of the study of microbes in ancient ice is rather short, but nonetheless, it is an interesting demonstration of how science really works. The scientists mentioned below are not well-known outside of scientific circles. Most are barely known on their own university campuses. But, to be clear, they all are intelligent, thoughtful, and insightful scientists. We have met many scientists who have been involved with this work. They are interesting and cordial people. In fact, if you had dinner with them, you would definitely have a good time. Some are hilarious, while others are dead serious. Some are the nicest people you would want to know, while others are a bit edgy. They are all unique characters, just as with any group of people you might meet. The first accurate written descriptions of ice and permafrost were published in the 18th century. During this and later times a variety of animals and plants were found in ice and permafrost. The discoveries of Woolly Mammoths often made the biggest headlines (Fig. 10.7). These enormous hairy relatives of elephants had been extinct for at least 4,000 years. During the 19th century further studies on ice and permafrost clearly showed that permafrost contained a diversity of living microbes. However, during that time they were unable to determine the age of the ice and so were unaware of its antiquity.

Bacteria that form hydrogen sulfide (the rotten egg smell, indicating the presence of bacteria that usually grow only in the absence of oxygen—anaerobes) and those that fix nitrogen (take nitrogen from the atmosphere and chemically bind it into compounds that are usable to other organisms, including humans) were isolated. This last group is very important, because these bacteria are able to take atmospheric nitrogen and convert it into compounds that can be used by plants and other organisms as a type of fertilizer. These reports were the first hints that some of the microbes in ice could be beneficial. By the late 19th and early 20th centuries, more detailed microbiological studies were underway, and the first calculations of concentrations were made. Sometimes tens of thousands of viable microbes were isolated from each milliliter of water melted from permafrost and glacial ice. Furthermore, more bacteria grew on media supplemented with meat extracts than with those supplemented with plant extracts, indicating that many might grow on animals, including humans, and therefore some might be pathogens. There were many expeditions to the Arctic and Antarctic at that time, and scientists were sent along with these expeditions to collect and catalog the plants, animals, and other organisms, including microorganisms. Thousands of new species of polar bacteria and fungi were described, and brought back to laboratories for study.

Most studies published in the mid-20th century were descriptions of microbes primarily of interest to other scientists, and thus they remained unknown to most people. One study examined some of the microbes isolated from the trunk of a Woolly Mammoth. Renewed interest in polar microbiology began in the 1970's and 1980's, when Professor Sabit Abyzov, in Moscow, Russia, began to publish articles describing a wide variety of microbes in glacial ice and ancient ice cores. By this time, methods to date the ice were more sophisticated and accurate. This was an important development, because Dr. Abyzov could conclude that the organisms being isolated from ice cores were thousands of years old. This grabbed the attention of scientists and non-scientists alike, although there were many skeptics. Nonetheless, Dr. Abyzov and his colleagues are clearly the pioneers of this area of study. His lab members built some of the first lab apparatuses for melting ice aseptically, so that the organisms inside the ice could be separated from those contaminating organisms on the outside surfaces of the ice. Although there have been improvements in this apparatus, forms of it are still used today, decades after its invention.

Another pioneer in ice biology in the 1970's and 1980's was Professor Wolf Vishniac (son of famed photographer Roman Vishniac). He and his wife, Dr. Helen Vishniac, extensively studied Antarctic microbes, and were among the first to suggest that parts of Antarctica could be used as models for parts of Mars. This modeling included testing of equipment and searching for life in extreme environments. Their initial ideas are finally coming to full fruition, because over the past decade, extensive areas of water ice have been found on Mars. The most massive zone of ice is in the north polar region of Mars. Therefore, the methods and equipment developed for examining life in ice on Earth (known as the "Wolf Trap") may very soon be used to seek life on Mars. Tragically, Professor Wolf Vishniac died during an expedition to Antarctica when he fell while in the field collecting samples, a stark reminder of the dangers of working in remote polar regions. Happily, his wife carried on their research, and he is remembered for his valuable contributions to the field.

The field of ancient biomolecules had its beginnings in the 1980's, prior to the advent of polymerase chain reaction (PCR) methods, which have increased the sensitivities for detection of DNA and RNA. Of course, at that time researchers (including one of us—SOR) were not thinking about specimens from ice. They (we) were using mummified plants and animals (including humans). In 1984, Dr. Russ Higuchi and coworkers published the DNA sequence of a gene fragment from a quagga, an extinct member of the horse family. The piece of muscle tissue that they used came from a museum specimen that had been cut from an animal that died a century earlier. This was an astonishing report at that time. Although degraded, the short pieces of DNA were cloned into a bacterial plasmid and sequenced. These sequences were then used to determine where the quagga belonged on the evolutionary tree of horses. It turned out to be in a place different than that determined using morphological characters. Although it had stripes reminiscent of zebras, the quagga was actually closer to horses than to any zebra species. This was a hint of the value of ancient DNA. After reading his paper (as soon as it was published), one of us (SOR), while a PhD student in the lab of Professor Arnie Bendich, attempted to isolate DNA from ancient plant specimens. To our delight and surprise, our attempts to extract and characterize DNA from herbarium specimens (preserved dried plants) to over 100 years old, and mummified specimens (from ancient packrat dens, called middens) over 45,000 years old were successful in almost every case. Although some samples yielded very degraded DNA, some (even the oldest) contained surprisingly long pieces of DNA. We were very skeptical of the initial results, since this was something that had never been reported, and we were conditioned to believe that dead organisms contained biomolecules that were so degraded that they would be useless. But after numerous experiments and controls, there was no longer any reason to doubt the finding that DNA was preserved for hundreds to at least tens-of-thousands (and possibly hundreds-of-thousands) of years in dried specimens. That research was published in 1985. In the same year, Dr. Svante Pääbo reported DNA that he had isolated from an Egyptian mummy. He has since formed an entire institute in Germany that pursues research in ancient biomolecules, primarily from mummified specimens. Also, in 1985, Dr. Franco Rollo reported the extraction of DNA from ancient plant seeds from the Necropolis in Thebes. He subsequently isolated DNA from corn kernels entombed in ancient Mayan structures. In the ensuing years, we attempted DNA extraction and characterization from other types of ancient specimens, including plants, fungi, algae, animals, and bacteria. Again, the majority were successful. All of these DNAs were thousand to tens-of-thousands of years old. Therefore, there was no doubt that DNA could be preserved for at least this long.

SERENDIPITY STRIKES

As with most scientific projects, conversation about something quite simple often develops into years of detailed projects that are not so simple. This was the case when Dr. Jack Fell and Dr. Tom Starmer (two of our colleagues) sat on a warm Caribbean island having a few drinks. Jack had spent decades describing hundreds of new species of yeasts from polar regions. Tom had spent a similar amount of time describing yeasts (and the fruit flies that depend on them) from around the world in warmer climes. Tom and Jack had worked together in Dr. Herman Pfaff's lab at the University of California, Davis, so they were old friends. They also had common research interests, and even after moving away from the Pfaff lab, they often worked together collecting fungi. In the late 1980's they were in the Caribbean studying these fungi, when they stopped to have a drink and a conversation about their research. As they sipped their drinks, Jack asked Tom about possible pathogens in the ice cubes in their drinks. After all, they didn't want to get sick. The unanimous answer was that there probably were quite a few viable microbes in the ice cubes, and that many could be human pathogens. Nevertheless, they continued to sip their drinks, knowing that often travelers to foreign countries contract diarrheal diseases from drinking the local water or eating ice. Then, the conversation shifted to microbes in glacial ice. The question was not "Are they there?", but "How many are there?" The short answer that would later be answered was, "A lot!" After this and subsequent conversations they initiated their first study of viable microbes in glacial ice. Ice cores from Greenland were available at that time from a storage facility in Buffalo, NY (but they were later moved to the National Science Foundation Ice Core Laboratory, NSF-ICL, previously the National Ice Core Laboratory, NICL, in Lakewood, Colorado). Tom was at nearby Syracuse University. He and others in his lab obtained some of the ice and began to isolate bacteria, fungi, and other organisms from the ice. An amazing diversity of culturable organisms emerged. This was startling, since only a small proportion (usually, around 1%) of the viable organisms are culturable, thus indicating a huge collection of microbial life was locked away in the ice.

We became involved shortly thereafter at the suggestion of Dr. Cathy Catranis (a graduate student at the time), who was working with Tom. At that time, they were using culturing, microscopy, and physiology methods to examine microbes from the ice, and they wanted someone with molecular experience to examine the microbes they had isolated. I (SOR) could do this, having developed methods to isolate DNA from small amounts of ancient and modern samples. Thus, most of the research team had been assembled. But, one group of microbes was missing from everyone's analysis. No one had looked for viruses in the ice. This was primarily due to the fact that no virologists were involved, but more importantly, viruses are difficult to find, and difficult to work with once found. Although they are more difficult to study, they are very important microbes. They cause a multitude of devastating plant and animal diseases, and have been important agents of evolution. JDC (who was only five doors down the hall from SOR) had worked with viruses for decades. So, JDC, SOR, and Tom met for the first time in the early 1990's. In the initial meeting each was surprised and excited by the findings and methods of the others. We had all worked within 30 to 300 yards of one another for between 6 and 15 years, but had not gotten together sooner to discuss our common research interests. Tom had the expertise in fungal methods and population studies, JDC had expertise with viruses and bacteria, and SOR had developed the necessary molecular methods for the isolation and characterizations of DNA from small amounts of ancient materials. After the initial meeting, we started planning, and the research began almost immediately. Our focus would be to determine sequences from the organisms already isolated at that time by Tom and Cathy, as well as to isolate more organisms from newly acquired ice cores, determine sequences directly from the ice, and to look for viruses. Important in the extension of this work were our colleagues, Dr. Li-Jun Ma (who isolated and characterized some of the fungi), Dr. J. Smith (who isolated and characterized bacteria and viruses), Virginia Aberdeen (who characterized some of the yeasts), and Yinghao Zhao (who characterized some of the bacteria and viruses).

The development of the field of studying ancient microbes in ice is similar to that of most fields of scientific study. Often it begins with a few very novel research reports that almost always go unnoticed, sometimes for decades or longer. Frequently, a group of scientists meet for dinner or a drink. One brings the idea to the table and the various pieces of the puzzle converge from the ensuing discussion. Alternatively, a scientist (often a graduate student) goes to the library, and in a moment of serendipity reads an intriguing paper and comes up with a new way to expand upon the idea. Frequently, this involves new laboratory techniques and/or reading historical publications. Once the ideas come to the forefront, one of the inevitable questions is, "Is this worth pursuing as a research project?" More often than not, the answer is "No." However, the answer is very dependent on who is asking the questions, since all scientists are unique and have different interests, expertise, and modes of operation. If the answer is "Yes," then some initial experiments are performed. The results are then used to secure additional funding to continue the work. This is done by submitting grant proposals, which generally take months to write. If luck and data are with you, funding comes upon a first submission. Not surprisingly, our proposal was declined in the first try, as are about 95% of first proposals. The reviewing panel decided to fund more mainstream research projects. But, on the second try, we were granted funds from the US National Science Foundation (NSF) to pursue the research.

Meanwhile, several other research groups began biological and physiological investigations into microbes in ice. Dr. David Gilichinsky and coworkers, from Russia, began isolating microbes from Siberian permafrost. The samples ranged in age from a few hundred years old to millions of years old. A few other laboratories also were involved, including Drs. John Priscu and David Karl, who isolated organisms from lakes and ice in Antarctica, and a collaborative group consisting of Drs. John Reeve, Lonnie Thompson, Ellen Mosley-Thompson, and Brent Christner (at that time a graduate student) who examined microbes from the Himalayas, the Andes, and Africa's Mt. Kilimanjaro. Again, most research focused on bacteria, although occasionally fungi and algae were recovered. At this time, the microbiologists, glaciologist, ecologists, and the molecular biologists had not yet integrated their research efforts. This is common in science. It is difficult to stay current in your own field, so it is seldom possible to also stay up to date in other fields at the same time. The researchers kept progressing in mutually exclusive directions, until the early 1990's, when several groups more-or-less simultaneously started characterizing both microbes and nucleic acids from ice and permafrost.

During this time, a 4,000-year-old mummified corpse was found by hikers in the Tyrolian Alps. The unfortunate man had been killed by one or more arrows and covered by snow that eventually became part of a glacier. Movement, plus recent recession of the glacier (due to climate change) eventually caused the man's body to fall out of the end of the glacier. The degree of preservation was astounding. He was transported to a laboratory in Italy and was then examined, described, poked, prodded, and tested. Microbes were grown from various parts of his body and from his clothing and shoes. A few years later, a frozen mummified corpse of a young Incan girl, who was killed in a ritual sacrifice, was found in the Andes mountains. Again, various tests were performed. This grabbed the attention of the media, public, and many scientists, including ourselves. Unfortunately, protection of the corpses from modern contaminating microbes was inadequate resulting in microbial examinations that were inconclusive. However, all of these reports were pointing in the same direction: Ice can preserve biological materials for very long periods of time.

In the 1990's a huge subglacial lake, as large as Lake Ontario, was discovered under 4 kilometers (approximately 2.5 miles) of Antarctic ice. Lake Vostok had been predicted by Andrey Kapitsa, a Russian researcher, decades earlier, but it was only confirmed and carefully mapped using seismic and radar methods in the 1990's, by a British research team, headed by Dr. J.P. Ridley. Some initially thought that the lake must be sterile, but this turned out not to be the case. Although only a small slice of the lake has been investigated so far, we found a diverse set of organisms in Lake Vostok. The lake receives ice meltwater from the overriding glacier, and that ice was deposited more than 400,000 years ago. The lake appears to have been continually covered by ice for at least 14 million years. Being covered by so much ice presents some big challenges to researchers interested in life in the lake. One way to sample microbes from the upper layers of water is to use accretion ice that has built up at the surface of the lake. Between the glacier and the lake, there is a layer of ice 200 m thick that has formed by lake water freezing to the bottom of the glacier. Thus, microbes from the lake have already been automatically sampled. Cores from this accretion ice have been assayed in several research laboratories. We performed extensive analyses on meltwater from the accretion ice from about 2000 to the present, using culture techniques, microscopy (light, fluorescence, and electron), and standard sequencing, as well as metagenomic and metatranscriptomic methods. From these studies, we determined that the lake contains thousands of species of organisms, most of which are unicellular bacteria, but a few small eukaryotes also were detected. Interestingly, some of the organisms are similar to those found in other subglacial lakes, while many others are unique to Lake Vostok. Details are presented in a later chapter.

For more than a decade, scientists had been planning and debating how and when to drill through the last remaining sections of Vostok accretion ice in order to sample the subglacial lake water directly. Punching through presented many challenges. The lake water is under tremendous pressure, due to the 4 km of overlying glacial ice. What would happen once the last of the ice was removed from the borehole? Also, the borehole was constantly filled with drilling fluids, containing high amounts of kerosene, that kept the borehole from freezing shut. Would the fluids contaminate the lake water? Additionally, several species of bacteria grow in the drilling fluids, so would they contaminate the lake? In 2012, Russian scientists finally got the approval to break through to the lake water. The theory was that the pressure in the lake water would push the water up into the borehole to a height of about 50–70 meters. They would then let the water freeze, and would come back the next season (2013) to drill into that frozen lake water, retrieve a core, and analyze it for the presence of drilling fluids, gases, and microbes. They drilled and retrieved a core in 2013, and began the analysis. To their disappointment, the core section was heavily contaminated with drilling fluids and the microbes that grow in the drilling fluids. They tried again in 2015, but the results were never published. Presumably, it was also contaminated.

In 2013, another subglacial lake, named Lake Whillans, was examined. It is smaller (covering only 60 km^2 versus 12,500 km^2 for Lake Vostok) and shallower (mean depth of 2 m versus 1,165 m for Lake Vostok), that lies beneath 800 m of glacial ice, near the Ross Ice Shelf in Antarctica. Water samples from the lake and a mud core from the lake bed were retrieved using a method that minimized any contamination from the surface. Sequences from several thousand species of microbes were recovered from the lake water and lake bed. The community of organisms indicated an ecosystem relying primarily on the oxidation of ammonia, iron, sulfur, and methane. In 2015, near the grounding line of the same glacier, a video camera that was lowered down a borehole, which had been drilled completely through the Ross Ice Shelf (740 m thick) into the ocean shoreline below. It revealed the presence of fish, crustaceans, and jellyfish. Some of the fish were almost 20 cm (7.5 inches) in length. The location was near the outlet stream from Lake Whillans, where its relatively fresh water mixed with the salt water of the ocean. This habitat is cold and in total darkness, yet appears to support not only microbes, but large multicellular organisms. The bottom line of all of these studies is that large complex ecosystems exist in icy environments around the world. Scientists were somewhat surprised by some of the results, because many of these sites were predicted to contain few organisms. The eventual results from these projects turned out to be much more interesting and complex than the initial theories of what existed under the ice.

SOURCES AND ADDITIONAL READINGS

Castello, J.D., and S.O. Rogers. 2005. *Life in Ancient Ice*. Princeton, NJ: Princeton University Press.

Castello, J.D., S.O. Rogers, W.T. Starmer, C. Catranis, L. Ma, G. Bachand, Y. Zhao and J.E. Smith. 1999. Detection of tomato mosaic tobamovirus RNA in ancient glacial ice. *Polar Biology* 22: 207-212.

Christner, B.C., J.C. Priscu, A.M. Achberger, C. Barbante, S.P. Carter, K. Christianson, A.B. Michaud, J.A. Mikucki, A.C. Mitchell, M.L. Skidmore, T.J. Vick-Majors, and the WISSARD Science Team. 2014. A microbial ecosystem beneath the West Antarctic ice sheet. *Nature* 512: 310-313.

Desmond, A., and J. Moore. 1991. *Darwin, The Life of a Tormented Evolutionist.* New York: W. W. Norton & Company.

D'Elia, T., R. Veerapaneni, and S.O. Rogers. 2008. Isolation of microbes from Lake Vostok accretion ice. *Appl. Environ. Microbiol.* 74: 4962-4965.

D'Elia, T, R. Veerapaneni, V. Theraisnathan, and S.O. Rogers. 2009. Isolation of fungi from Lake Vostok accretion ice. *Mycologia* 101: 751-763.

Ma, L., C. Catranis, W.T. Starmer, and S.O. Rogers. 1999. Revival and characterization of fungi from ancient polar ice. *Mycologist*, 13:70-73.

Ma, L., S.O. Rogers, C. Catranis, and W.T. Starmer. 2000. Detection and characterization of ancient fungi entrapped in glacial ice. Mycologia 92: 286-295.

Rogers, S.O. 1994. Phylogenetic and taxonomic information from herbarium and mummified DNA. In: R.P. Adams, J. Miller, E. Golenberg, and J.E. Adams (eds.) Conservation of Plant Genes II: Utilization of Ancient and Modern DNA. Missouri Botanical Gardens Press, St. Louis, pp. 47-67.

Rogers, S. O. 2017. *Integrated Molecular Evolution*, 2nd ed. Boca Raton, FL: Taylor & Francis Group.

Rogers, S.O., Y.M. Shtarkman, Z.A. Koçer, R. Edgar, R. Veerapaneni, and T. D'Elia. 2013. Ecology of subglacial Lake Vostok (Antarctica), based on metagenomic/metatranscriptomic analyses of accretion ice. *Biology* 2: 629-650.

Rogers, S.O., Theraisnathan, V., Ma, L.J., Zhao, Y., Zhang, G., Shin, S.-G., Castello, J.D., and Starmer, W.T. 2004. Comparisons of protocols to decontaminate environmental ice samples for biological and molecular examinations. *Appl. Environ. Microbiol.* 70: 2540-44.

Shtarkman Y.M., Z.A. Koçer, R. Edgar R, R.S. Veerapaneni, T. D'Elia, P.F. Morris, and S.O. Rogers. 2013. Subglacial Lake Vostok (Antarctica) accretion ice contains a diverse set of sequences from aquatic, marine and sediment-inhabiting Bacteria and Eukarya. PLoS ONE 8(7): e67221. doi:10.1371/journal.pone.0067221.

LINKS

https://en.wikipedia.org/wiki/%C3%96tzi
https://en.wikipedia.org/wiki/Ancient_DNA
https://en.wikipedia.org/wiki/Arctic_exploration
https://en.wikipedia.org/wiki/Charles_Darwin
https://en.wikipedia.org/wiki/Genetics
https://en.wikipedia.org/wiki/Gregor_Mendel
https://en.wikipedia.org/wiki/History_of_Antarctica
https://en.wikipedia.org/wiki/International_Geophysical_Year

11 Ice Core Discoveries

"What we see depends mainly on what we look for"

John Lubbock

ALL ICE IS NOT THE SAME

Many assume that all ice is much the same. Also, when they think of liquid water, they think it is all the same. However, just as water can contain impurities (including microbes) and dissolved chemicals, so, too, can ice. Think of drinking water out of a clear mountain stream and then think of drinking water from the ocean. In addition to one being clear and clean-tasting, and the other very salty, both contain hundreds to thousands of microbes per drop, and they have the potential to contain some pathogenic organisms, some of which can kill you. This is also true for ice. A piece of ice from a glacier is not necessarily safe to eat.

Most people learn that water freezes at 0°C (32°F) and boils at 100°C (212°F). However, very pure water, as well as water containing certain chemicals, can remain liquid well below 0°C. This is super-cooled water and can immediately freeze if a bit of energy (such as a vibration) or a small particle (e.g., sand grain, small ice crystal, microbe) is introduced. When this happens, very small crystals form. Organisms that survive in ice often produce cell components (often proteins) that will cause ice nucleation, and which allow only very small ice crystals to form that are too small to damage the cells. When ice freezes near 0°C, very large crystals can result, because the water molecules are added to the growing crystal in a more organized way. As large crystals grow, they slice through cells, killing them. Also, if ice forms slowly, it forces solutes out of the crystal such that large monocrystals of pure water ice can result, and where the crystals meet, solutes concentrate, resulting in narrow veins of liquid water.

IGY

Much of the study of environmental ice, especially in the Arctic and Antarctic began in the middle of the 20th century. The year 1957/58 was designated the International Geophysical Year (IGY). Many polar stations, including US stations, Byrd, McMurdo, and Amundsen-Scott (at the South Pole), in Antarctica (US), were established during the planning and build up to the IGY. At about the same time, stations also were built by France, Australia, Chile, Argentina, Poland, Soviet Union (Russia), United Kingdom, New Zealand, Belgium, Norway, Chile, South Africa, and Japan. Polar exploration and study expanded from that point, and has continued to grow through the ensuing years. Much of the early research centered around descriptions of glaciers, ice depths, characterization of the dry valleys in Antarctica, and others. The first cores drilled in the 1960's revealed that stable isotopes of various elements within the trapped gases could be used to determine the age of the ice, as well as past climate changes. Subsequently, more drilling methods were developed, and more ice core projects and studies were initiated and accomplished. Eventually, biologists studied the flora, fauna, and microbes of the ice cores, as well as from a large number of polar sites.

Drilling became more sophisticated through the utilization of drilling methods developed for geological studies to determine the locations of oil and gas deposits. Several methods were developed and tested at various sites in Greenland and Antarctica. Several types of mechanical drills have been used, including those that used no drilling fluids (so-called, "dry drilling"), and those that used various types of fluids, including several different types of hydrocarbon liquids such as kerosene, freon, and other liquids, that aid in lubrication of the drill, reduction of vibration, facilitation of ice chip removal during drilling, and prevention of refreezing of the borehole. These drills have a ring of cutting teeth around the perimeter of the drill, and the core section is the ice that is drawn up into the center of the drill body (see Fig. 8.1). Usually, 1–3 m (3–10 ft) core sections (depending on drill type) are brought to the surface at a time. Thermal drills have a ring-shaped heating element that takes the place of the cutting teeth in a mechanical drill. As with the mechanical drill, the core section ends up in the center of the body of the drill. Hot water drills have been used when a specific depth of core, or a lake, is to be sampled. Hot water drills shoot high pressure hot water out of their end nozzles, and cut rapidly through the ice. Once the desired depth is reached, a mechanical or thermal drill is used to obtain the desired core, or to cut through to a lake.

1950'S AND 1960'S

Several ice core projects were initiated during the late 1950's and 1960's. Drilling at Site 2 (near Thule Air Base and Camp Century, Greenland, Figs. 1.2 and 8.5) began in 1956, where two cores (305 m and 411 m in depth; 1,000 to 1,350 ft) were drilled. In 1957–1959, ice cores at Byrd station (Antarctica) were drilled, again resulting in relatively shallow cores. However, measurements of gases and isotopes in the various sections of these ice cores led scientists to understand that accurate measurements of conditions on Earth long into the past could be made by analyzing gases and isotopes in the ice core sections. While the ice at Byrd was drilled only to a depth of 309 m (1,014 ft) at that time, by 1968, drilling had progressed to bedrock at 2,164 m (7,100 ft).

Drilling at Camp Century, Greenland, was initiated in 1960 (see Fig. 1.2 and 8.5). The site was actually created as a military base from which to launch missiles towards targets in the Soviet Union. Not only did the IGY stimulate an increase in activity to study polar ice, but it initiated an increased interest to place radar and other offensive and defensive posts in polar regions. The fact that the Camp

Century drilling project was initially a cover for a military purpose was kept secret from Denmark (Greenland is part of Denmark) until 1995! Happily, the site became more useful as a scientific endeavor than a military one because of instability of the ice in the region, which made missile silos impossible to construct. However, while no nuclear missiles ended up at the site, planes at nearby Thule airbase did carry nuclear weapons, including several that were in a plane that crashed on the ice. While most of the nuclear devices were recovered, portions of one nuclear bomb went deep into the ice and were never found.

The ice core sections collected in the Camp Century project demonstrated that atmospheric gases trapped in the ice could be used to determine past gas concentrations. Stable isotopes also could be used to determine past temperature and climate changes, which was later confirmed in other ice core projects. The core sections collected at Camp Century eventually allowed researchers to reconstruct past climates back to 13,000 years ago. The Medieval warming period from 800–1200 CE, a local warming event centered around Greenland, Iceland, and Western Europe was documented. It also documented a gradual cooling period that led to the "Little Ice Age" beginning around 1,300 CE, and continuing through the middle of the 19th century. These results were consistent with those from other core studies (e.g., Milcent and Crete). During cold times, ice crystals are small, while during warmer periods, larger ice crystals form in the ice fields and glaciers. More dust is deposited in ice that accumulated in cold periods, while less dust is deposited in ice from warmer periods. This occurs because there is less moisture in the atmosphere during cold periods, which makes the land dry, leading to more dust blown into the air. More microbes are found in the dust-containing ice because the winds also carry the microbes, either as individual cells, mats of cells, or cells on and in dust particles. Many of the microbes in this ice are viable. In warmer periods, more moisture is in the air, and therefore, there are more frequent periods of precipitation that washes the dust and microbes from the atmosphere. The land is wetter, which leads to less dust blowing into the atmosphere. These results are consistent for ice collected in Greenland, Antarctica, the Himalayas, and elsewhere.

Some early ice cores were used to examine events that occurred within the past few thousand years, partly because the cores were relatively shallow (i.e., they represented a short timespan), and partly because the core section dates could be correlated with recorded historical events and results from ocean sediment cores. In particular, volcanic eruptions that produced large amounts of atmospheric dust and sulfates that were deposited in the ice, were informative, in that they could be accurately dated. Dust can cause decreases in the amount of sunlight reaching the surface of the Earth, and thus can negatively affect agriculture. Sulfates in the atmosphere reflect sunlight, and therefore also negatively affect agriculture. Each instance of a large deposit of dust and sulfates in the ice, indicating a large volcanic eruption, correlated with decreases in summer temperatures around the world, and many reports of crop failures, famines, and disease epidemics. So, the inclusions frozen in ice cores can say a lot about many events that have occurred in Earth history, far into prehistoric times. One GISP 2 core section that we (with Caitlin Knowlton, then a graduate student in my lab) analyzed was dated to 10,500 years before the present. In addition to having sequences from many bacteria and fungi, it contained particulate matter. Elemental analysis of these small grains indicated that they probably had come from a volcanic eruption. At the time, there was geologic evidence for volcanic activity in Alaska, Iceland, and Italy. But, in addition to elements typical of many volcanoes (e.g., silicon, iron, and magnesium), the particles in the 10,500-year-old GISP 2 ice core section also contained elevated amounts of tungsten, which is rare in volcanic ash. However, a few volcanoes in Alaska do have moderate concentrations of tungsten, which pointed to a volcanic eruption in Alaska during that time. Some sequences from that ice core section were from species of thermophilic bacteria. Presumably, they had been residents of the soils or hot pools on the volcano when it erupted. They may have been blown high into the atmosphere, and then were eventually deposited on the Greenland glacier, possibly in snow.

1970'S

In the 1970's, most research on past climate events was being done using sediment samples from ocean beds. They also provided information regarding organisms, because microfossils, pollen, and even living microbes could be recovered from some sediment cores. It was during this decade that ice core data were being compared with the ocean sediment core data, and the correlations were abundantly clear. The warming and cooling of the Earth during the past centuries and millennia were synchronous in all of the data sets. Shallow and intermediate depth ice cores were drilled and collected from the Dye 2 and Dye 3 sites (Figs. 1.2 and 8.5), as well as a deep core from Dye 3 (where bedrock eventually was reached at 2,038 m, 6,686 ft, in the early 1980's), to document these warming and cooling periods. Dye 2 and Dye 3 were originally part of the Distant Early Warning (DEW) line, begun in 1957, as part of a system to warn of air attacks (airplanes or missiles), and was connected to the North American Aerospace Defense Command (NORAD). As with Camp Century, the sites became used partly for ice core projects. Dye 2 has since been decommissioned. Dye 3 is still used for military training. While much of the Dye 3 core produced very good sections, brittle ice was encountered between 600 and 1,200 m (1,970 to 3,937 ft), which produced information of poor quality and continuity. Below 1,600 m (5,250 ft) the ice again was of good quality. At the bottom, the ice was approximately 110,000 years old. The date is important because the Earth was emerging from its most recent interglacial period, called the Eemian period, that lasted from about 130,000 to 115,000 years ago.

Another important ice-coring project was initiated in the 1970's in Antarctica. The site was called Vostok (Fig. 1.2), a remote, high-elevation, extremely cold location, about 1,300 km (800 mi) from the South Pole. Although it was established in 1957 by the Soviet Union, coring only started in 1970. Larger drilling projects began in the 1980's. While there is 24-hour sunlight during the brief summer, there is total darkness during the winter. Even in summer, it is bitterly cold. This is not a hospitable place. A few dozen scientists and support staff are present in the summer, but fewer than a dozen support staff remain for the winter, primarily to keep all of the equipment in running condition. To demonstrate how hazardous it is to work in places like Vostok and the South Pole, if anyone becomes dangerously ill during the winter, the chances of being evacuated to a hospital and medical help are very low. Because of the distances and altitudes, only airplanes can reach these sites. The air is too thin for helicopters. Also, the air is so cold that the hydraulic fluids in airplanes can freeze, making flights into and out of these areas impossible, or at least hazardous, for most of the winter. During one summer, the main generator at Vostok caught fire, and was badly damaged. All activities at the station were suspended, and people were evacuated until repairs were completed. Survival at the station is impossible without the generator. Luckily, the fire did not occur during the winter, which could have been fatal.

Vostok station lies in a cold desert, with an average of 22 mm (0.87 inches) of snowfall annually. This means that the yearly snow and ice layers are very thin, such that a section of core encompasses a long period of time. Thus, the cores at Vostok represent hundreds of thousands of years of past snowfall and atmospheric gases. The earliest cores reached depths of 500 m (1,640 ft) and less. But, because of the low ice accumulation rates, this was relatively old ice. However, often the drills got stuck, and some had to be abandoned. So, the Russian drillers attempted a new technique (at that time), called deviation drilling, where they started a new core by drilling at an angle in the previously drilled borehole to go around the trouble spot. Ultimately, they drilled three major cores, termed Vostok 3G, Vostok 4G, and Vostok 5G (Fig. 8.5), utilizing this method in each of the boreholes.

1980'S

From 1980 through 1985, drilling continued on the Vostok 3G core, first getting stuck at 1,590 m (5,217 ft), and then sticking a second drill at 2,085 m (6,840 ft). However, at that depth, the ice was more than 150,000 years old, and covered the last major ice age, as well as the interglacial period (Eemian period, that preceded the present interglacial period in which we live). As part of the Vostok project, Professor Sabit Abyzov began the foundational microbiological studies of the ice core sections. He documented the presence of many different bacteria, using light and fluorescence microscopy. He set the wheels in motion for microbiological research of ice cores. Some ice core sections contained an abundant and diverse assemblage of microorganisms, whereas other sections (sometimes adjacent ones) contained virtually nothing. Abyzov and coworkers have used direct microscopic examination of filtered ice meltwater, and the consumption of radioactively labeled organic compounds as proof of viability rather than direct culturing on nutrient media. He was able to show that ice cores up to 240,000 years old obtained from Vostok, Antarctica, contained a wide variety of contemporary bacteria, yeasts, filamentous fungi, microalgae, pollen of higher plants, and dust particles. The filters then were stained with a fluorescent dye, which complexes with proteins, and examined by epifluorescence microscopy to quantify the number of cells present in the meltwater. SEM (scanning electron microscopy) was conducted on the filters as well to observe the size and shape of the cells. Vegetative bacterial cells, but not endospores, were detected. In addition, the number of bacteria detected was directly related to the amount of dust present in the ice, which varied by depth. In 1985, a new core, Vostok 4G, was initiated. This core was blocked at only 279 m, (915 ft) and deviation drilling again was performed. However, this was blocked at 751 m (2,464 ft) by the summer of 1986. Again, they were forced to perform deviation drilling, and in this attempt they continued drilling to a depth of 2,546 m (8,353 ft) by 1990. At this depth, the ice was more than 240,000 years old, and the analysis revealed a great deal about the state of the Earth's atmosphere and temperature over that period. Microbial cell counts ranged from 800–10,000 cells/ml of meltwater. Microbial viability decreased with increasing depth, but viable organisms were found at a depth corresponding to 240,000 ybp. This covered most of the previous interglacial period, including all of the gas and temperature fluctuations. Essentially, this also covered the entire time of the existence of *Homo sapiens*.

1990'S

By 1990, most of the information regarding microbes living in polar regions, including those in glaciers, ice domes, lake ice, and sea ice consisted of descriptions of the organisms; including their morphology, physiology, ecology, and similar features. However, measurements of gas concentrations and temperatures in the past were being compiled. This provided a great deal of ground work to begin to define cold polar ecosystems at the microbial level. By the 1990's, molecular methods were being used to augment the morphological and physiological methods to characterize the microbes and microbial communities. Initially, PCR, coupled with standard Sanger sequencing methods were used. But, by the late 1990's and early 2000's, the first of the next generation sequencing (NGS) methods became available. However, initially they were prohibitively expensive for most labs, and the sequences obtained were very short, so their use was limited. But, within a span of 10 years during the 2000's, the costs began to decrease rapidly, and the sequence lengths increased. Ultimately, these methods led to metagenomic and metatranscriptomic projects to determine which organisms and biochemical pathways were present in these samples. This led to a more thorough understanding of the types of organisms that exist in these cold environments, and some of the adaptations that they possess that allow them to survive and thrive in these locales.

The US-sponsored Greenland Ice Sheet Project (GISP), which started in the 1970's with the Dye 2 and Dye 3 ice coring sites, and the European-sponsored Greenland Ice Core Project (GRIP) began coordinated efforts in the 1980's to obtain and study ice cores that covered more than 100,000 years of ice accumulation (Figs. 1.2 and 8.5). In doing so, they produced a climatic record unprecedented in its detail. The cores were more than 3,000 m (9,843 ft) in depth and stretched back more than 105,000 years. While we were studying organisms from the Dye 3 ice core, we also studied microbes from GISP 2, which was part of the large study of ice in Greenland, called the Greenland Ice Sheet Project. This core was drilled from the late 1980's through the early 1990's, reaching bedrock in 1993 at 3,053 m (10,016 ft). At this depth the ice is more than 230,000 years old. This spans two ice ages and three interglacial periods (including the present one). Once the data were collected and compared with the Dye 3 and GRIP data, synchronized patterns were apparent. There were major cycles of warming and cooling with periodicities of 100,000 years, and others with shorter periods. The results documented the cooler drier period from 13,000 to 11,000 years ago, coinciding with the last ice age, and indicated that atmospheric carbon dioxide and methane increases coincided with warming periods, and lower concentrations of both coincided with cooler periods.

We examined ice from the Dye 3 and GISP 2D cores (with our collaborators, Tom Starmer, Cathy Catranis, Li-Jun Ma, Jack Fell, Virginia Aberdeen, and Yinghao Zhao) and found bacteria, fungi, and viruses in most of the core sections. Many of the bacteria and fungi were viable, but we could not determine if the viruses were viable. However, for one species of bacteria, *Bacillus subtilis*, we were able to induce a provirus (called PBSY virus) to enter its lytic phase, and produce new virus particles, indicating that the lysogenic viruses, whose genomes had been integrated into the host bacterial chromosome were still capable of producing new viruses, even though some of them had resided in the chromosome of the bacteria frozen within the ice for up to 110,000 years. Several key findings came from these initial studies of microbes in this ice core. First, viable organisms can be preserved in ice for more than 100,000 years. Second, these organisms could be cultured. Third, viability dropped with depth (age, see Fig. 3.3). Fourth, remnants of dead organisms can be preserved for more than 100,000 years. And, fifth, the genomes of certain species coming from different time periods appear to mix. We have termed this, "Genome Recycling," which we discuss in detail in a later chapter.

While we were examining the microbes in the Dye 3 and GISP 2 ice core sections, the Russian research group headed by Professor David Gilichinsky was performing similar analyses (primarily cultural and microscopy studies) on permafrost samples from Siberia. The fungal and bacterial species that they identified and isolated were similar to those we found in the Greenland ice core sections. Their permafrost samples were from 5,000 to 3,000,000 years old, while our ice core samples were from 300 to 110,000 years old. This implies a long-standing microbial ecosystem operates in the Arctic. Specifically, the species most often identified from both permafrost and ice cores from the north polar regions were: bacterial species from the genera, *Bacillus*, *Cyanobacteria*, and *Rhodococcus*; and fungal species from the genera, *Alternaria*, *Aspergillus*, *Cladosporium*, *Cryptococcus*, and *Rhodotorula*. Each of the genera have many species known to be adaptable to many environmental conditions. Some of them also include pathogenic species of plants, animals, and humans.

Viruses have been less studied primarily because they are more difficult to isolate and characterize, and the sequence databases have fewer virus sequences (especially for environmental viruses) than for bacteria and other microbes. We subjected several ice core sections from the GISP 2D and Dye 3 ice cores to analysis for tomato mosaic virus (ToMV) sequences. This virus infects a large number of plant species, not just tomatoes. Plant-pathogenic viruses often are named for the plant from which they were first isolated, in this case tomato plants. However, studies have shown that it infects many crop plants, as well as a large number of wild angiosperms (flowering plants) and gymnosperms (firs, pines, spruce, etc.). It is a widespread and hardy virus. Its relative, tobacco mosaic virus (TMV), can survive the burning of tobacco leaves (in cigars and cigarettes), and the ash remains infectious.

The sequences from the Greenland ice cores indicated a great deal of genetic variation for ToMV, which is characteristic of RNA viruses. In general, the sequences from more recent ice core sections (<5,000 years old) were very similar. In some of the older ice core sections, there were more nucleotide changes compared to sequences from more recent ice core sections, as well as from isolates of ToMV collected from living plants. However, some sequences from the oldest ice core sections, including one that was more than 110,000 years old, were almost identical to the sequences from modern ToMV isolates. This confused us at first, because we thought that the mutational changes would be neatly separated in time among the ice core sections. Eventually, we thought through the possibilities, and the most likely scenario was that the viruses, including their genomes, were being recycled in the ice. Put simply, some of the viruses at one point in time were frozen in the ice, while other viruses of the same species continued to infect plants, and this went on for hundreds of thousands of years or longer. As the organisms grew and multiplied, mutations occurred in their genomes, so over time, their genomes changed ever so slightly. At a later time, the frozen viruses melted out and mixed with the population of viruses that had been growing, multiplying, and mutating. The new population then contained a mixture of recent and older genomes. We have termed this, "Genome Recycling," which is discussed in more detail in a later chapter.

Researchers also have detected both marine and nonmarine diatoms (photosynthetic protists with silica shells) throughout ice cores from the South Pole spanning the past 2,000 years. They cut and melted the ice cores, and then filtered the meltwaters through 1.2-micron membranes, which were then dried, cut, mounted on slides, and examined at 1,000X to identify and count diatoms. Similar diatoms were found at drill sites in East and West Antarctica, at Siple and Taylor domes, and because there are no local diatom sources at any of these sites, the diatoms must have been carried by the wind from coastal or more distant sites. Diatoms are a small but persistent constituent of falling snow in Antarctica, although they occur in a patchy distribution in both space and time. More than 40 species of marine and nonmarine diatoms have been reported. Several genera were recovered from more than one age stratum. There are significant differences in diatom abundance by age stratum, ranging from 0 to 450 specimens per liter. Storms in

Antarctica tend to skirt the continent because of strong surface winds that flow from the center of the continent toward the sea. Occasionally, however, particularly strong storms can penetrate onto the continent itself. Because diatoms are light and easily carried by wind, it is likely that they are carried to the interior of Antarctica with these storms. Therefore, diatom abundance may indicate the historical frequency of these severe storms. To date, all of the information on diatoms is from studies of their shells (cell walls). No diatoms have been cultured from the ice cores.

In 1992, the drilling of the Vostok 5G ice core (Fig. 8.5) reached a depth of 2,500 m when the drill became stuck. They were still more than 1,000 m (3,280 ft) from the base of the glacier, and subglacial Lake Vostok, which was below the bottom of the glacier. So, again, they drilled around the blockage, and further into the glacier. By 1998, they had reached the 3,623 m (11,886 ft) mark. Examination of the ice core sections revealed that through 3,310 m (10,860 ft), representing a record of 420,000 years, the ice was in layers that could be dated and analyzed for gases and other components. So, there was now a record of carbon dioxide, methane, temperature, microbes, and other information extending back 420,000 years. There had been four ice ages and four interglacial periods during that time, including the current one. There is also a correlation between high atmospheric carbon dioxide, high methane levels, and low dust levels, in warm periods; and the opposite for cool periods. At no time during the past 420,000 years had atmospheric carbon dioxide levels exceeded 300 parts per million (ppm) in the atmosphere, yet the levels in 1999 had already gone above 300 ppm, and by 2016, they surpassed 400 ppm, and continue to rise. The ice below the 3,310 m (10,860 ft) depth was broken, disturbed, and twisted. This part of the glacier had interacted with the bedrock below the glacier as it moved downhill, and was broken and tumbled by those interactions. This "basal ice" often contains a different mix of minerals and organisms, and the microorganisms and ice are distorted. Although it is impossible to determine an exact age for much of this ice, some have estimated that portions are more than one million years old. Below the basal ice, starting at a depth of 3,538 m (11,607 ft) from the surface, very different ice, originating from frozen Lake Vostok water, called "accretion ice," was detected. Accretion ice is found above subglacial lakes, as well as on the bottom of sea ice and ice shelves. The Lake Vostok accretion ice holds a record of the lake water dating back more than ten thousand years. Details of this ice are discussed in a subsequent chapter.

In 1995, drilling started at Dome C, Antarctica, as part of the European Project for Ice Coring in Antarctica (EPICA, Figs 1.2 and 8.5). This project was focused on retrieving the longest continuous ice core record for any ice core. By 2000, they reached a depth of 2,864 m (9,396 ft), which was 500,000 years old. In 2005, they reached bedrock at 3,270 m (10,728 ft), where the ice was 800,000 years old, which was almost twice as old as the oldest ice from the Vostok ice core. When the results from the Dome C and Vostok ice cores were compared, they were in complete synchrony with respect to their profiles for temperature, carbon dioxide and methane levels, and amounts of dust. This included eight ice ages and nine interglacial periods, not all of which followed exactly the same patterns. The timing of each varied somewhat, and the fluctuations of gases and temperatures also differed. However, for the entire 800,000-year period, the atmospheric carbon dioxide levels never exceeded 300 ppm, except in the present interglacial period, where carbon dioxide levels have now surpassed 400 ppm.

Through the 1980's and the 1990's, several research groups also were studying ice from several mountain ranges throughout the world, including the Himalayas, sometimes called the "Third Pole" because of similarities to polar regions. In particular, Professor Lonnie Thompson collected ice from the Himalayas, the Andes, Mt. Kilimanjaro, and others. He pioneered the use of a solar-powered drill, which was developed so that climbing in the mountainous terrain was not hindered by having to carry the extra weight of heavy petroleum-fueled engines, and the fuel needed to power them. The solar-powered drill was lighter, and required no fuel to be carried up the mountains. These studies documented the recession of the glaciers around the world, and the relatively recent large changes in climate. His research also documented the El Niño and La Niña events in the Pacific, and how much the climate had changed over the past century. Climate change and global warming were clearly evident in the results.

2000'S

We, and our colleagues (Ram Veerapaneni, Tom D'Elia, and Vincent Theraisnathan), as well as others in the US and Russia, began our analyses of the biological contents of the Vostok ice core sections using light, fluorescence (using a stain that indicated whether the organism was living or dead), and electron microscopy, as well as culture methods. Very few cells and organic molecules were present in the glacial and basal ice, which was not surprising given the extreme environment of the area. Additionally, the number of viable microbes decreased with ice core depth (i.e., age, see Fig. 3.3). The number of viable organisms was especially low in the basal ice. The cells in the basal ice showed signs of pressure and damage when examined microscopically. However, we did revive three species (two bacteria and one fungus) through culture methods from the basal ice, which was estimated to be possibly more than one million years old. The ice below the basal ice was accretion ice, which is frozen Lake Vostok water. In the accretion ice, the ice and its contents were dramatically different. This accretion ice is remarkably uniform and dense, and is of two different types of ice; one containing a great deal of very fine particulate matter (type 1), and ice that was extremely clear, and almost completely devoid of particulate matter (type 2). Also, the numbers and diversity of organisms, the amount of biomass and organic molecules, and the number of viable organisms, as well as sequences obtained, increased dramatically in accretion compared to the basal ice (Fig. 3.3). These were the first signs that Lake Vostok probably was not sterile, but contained a diversity of life. We discuss this further and in detail in a subsequent chapter.

Starting in 2006, the WAIS (West Antarctica Ice Sheet) Divide project drilled and studied an ice core, culminating in 2011 with a 3,405 m (11,171 ft) ice core, estimated to be 68,000 years old at the bottom (Figs. 1.2 and 8.5). The location was on a dome area, which was chosen because the ice had been laid down in even sheets, and the ice was relatively static and stable. The data collected from the ice core sections produced a detailed view of the gases, temperature, and other characteristics of the ice over the 68,000-year timespan, showing that there were even more fine-scale fluctuations in each of the characteristics that were measured. However, it also was synchronous with the results from other Arctic and Antarctic ice core studies.

The NEEM (North Greenland Eemian Ice Drilling) project was begun in 2007 on the northern section of the Greenland ice sheet (see Fig. 1.2). The aim of the project was to obtain a complete record of the last major interglacial (i.e., warm) period, called the Eemian period, which occurred between 130,000 and 115,000 years ago. By 2010, the bottom of the ice had been reached, at a depth of 2,537 m (8323 ft), and an estimated age of 128,000 years old. Currently, we exist within the interglacial period that occurred after the Eemian interglacial period, so it was of interest to determine the conditions at that time to compare with current world conditions. The conditions during that period were similar to conditions that existed hundreds of years ago, but recent changes in atmospheric gases and global climate changes differ significantly today from what was measured for the Eemian interglacial period. The NEEM results were in synchrony with the results from other ice core studies with respect to temperatures and gases, but provided greater detail. One can conclude from this that the current extreme concentrations are caused by human activities, primarily the burning of fossil fuels.

Dome Fuji (Figs. 1.2 and 8.5) was established in 1995 by the Japanese, and the initial core was 2,503 m (8 212 ft) deep, with an estimated age of 340,000 years. A second core was begun in 2003, and by 2005, the depth of 3,260 m (10 696 ft) was reached, with an age of 720,000 years. The ice at this depth contained rocks and stones, as well as refrozen water, although this was not quite at the bedrock. This indicated some interaction of the ice with the bedrock, including some melting and refreezing. When compared to the EPICA results with respect to profiles of temperature, carbon dioxide, methane, and dust amounts, the two were synchronous throughout the 720,000-year period. In fact, when Fuji Dome, EPICA, Vostok, and oceanographic results were compared, they all lined up perfectly. Again, all agreed that the carbon dioxide levels were never above 300 ppm over the past 800,000 years. The only time the carbon dioxide levels have exceeded 300 ppm has been during the past two decades. Remember, *Homo sapiens* has existed only during the past 200,000 years, so this is also the first time in the history of our species that atmospheric carbon dioxide (as well as methane) levels have been this high. It will be challenging for humans to adapt to the abrupt changes in global climate, including changes in temperature, precipitation, storm severity/frequency, and sea level.

In 2007, Professor Eske Willerslev and his research team carefully identified plant species in the bedrock samples from the NGRIP ice core. Much of Greenland was, in fact, green about 450,000 to 800,000 years ago. Their results indicated that it was covered with plants that are characteristic of temperate forests during that time. But, starting about 450,000 years ago, the ice fields and glaciers began to expand, eventually almost completely covering Greenland. It is unclear what caused this abrupt transition at that time, but some clues may exist in the ice core data.

Ice core discoveries confirm that there are regular cycles, as well as irregular fluctuations, of the hot and cold times on Earth, some of which may have caused Greenland to transition from a temperate forest to an ice-covered landscape. During the past several hundred thousand years, the interglacial periods have occurred about every 100,000 years, but previous to that, they occurred about every 40,000 years. There also are other minor heating and cooling periods. One of them has a cycling time of about 20,000 years. These have been ascribed to the wobble of the Earth's axis, eccentricities in Earth's orbit around the sun, and passage through cosmic dust clouds. Also, there are irregularly occurring heating and cooling times. We are currently in one of those, a time of increases in carbon dioxide, methane, and global heating. It is clear that this perturbation of the normal cycles has been caused by human activities (i.e., pollution). But, other cooling times that lasted for months or years have been caused by past large volcanic eruptions (because of the dust and sulfates). Other climate events have been caused by releases of other gases and changes in ocean currents. Another cause of cooling has been from large asteroid strikes, at least some of which have caused global cooling that lasted for decades. One that struck Greenland about 12,000 years ago, caused widespread melting of the Greenland ice sheet. This event caused significant cooling, as well as salinity changes in the local ocean, leading to colder temperatures that may have persisted for a long period of time. A comet strike in 536 CE caused some of the coldest worldwide temperatures of the past 2000 years. They persisted for two to three years, causing crop failure, famine, disease, and death, not only for humans, but for plants, animals, and probably for microbes, as well. These drops in global temperature are recorded in ice cores from around the world.

2010'S

In the 2000's many new sequencing methods were developed and commercialized. These NGS methods allowed huge numbers of sequences to be determined. So, we and our colleagues (Zeynep Koçer, Yury Shtarkman, Ram Veerapaneni, Tom D'Elia, Robyn Edgar, Colby Gura, and Paul Morris) used NGS methods to perform several metagenomic (sequencing the DNA in a sample) and metatranscriptomic (sequencing the RNA in a sample) analyses of the glacial, basal, and accretion ice sections from the Vostok ice core. These studies confirmed that contents of the glacial and basal ice differed from that in the accretion ice. Additionally, the accretion ice that had originated from a shallow embayment in Lake Vostok contained the highest number and diversity of organisms, including some multicellular organisms. It also contained the highest concentrations of viable microbes. Some of the glacial ice contained very few sequences, and the basal ice also contained few sequences, and almost no viable cells. As with our studies based on microscopy and culturing, the metagenomic and metatranscriptomic analyses indicated that Lake Vostok contains a diverse set of living organisms, including some sequences being deposited by the glacier, and others being unique to the lake. The Lake Vostok results are detailed in a separate chapter.

In 1998, drilling at the Vostok site was halted just over 140 m (459 ft) from the surface of Lake Vostok. Then, for more than a decade, many meetings were held and many ideas, concerns, and proposals were voiced. The concern was how to get a sample of the lake water, or deliver a robot to the lake, without contaminating the lake itself. If the lake became contaminated, it would compromise further study of the lake, and might potentially change the lake forever. Several potential methods and robots were proposed, and one by one they were discussed and dismissed. Another major problem was that the borehole was filled with drilling fluids (primarily kerosene), to prevent it from freezing shut. A surprisingly large number of microbes live in the drilling fluid. They actually use the fluids as food sources for their metabolism, and they can grow and multiply in the fluid. Furthermore, it was impossible to filter them out of the fluid or add chemicals to kill them (which might also kill the organisms in the lake water). So, if the drill punched through to the lake water, the fluids were likely to contaminate it. However, the lake water is under high pressure, so once the drill punched through to the lake, it was assumed that the water would rise in the borehole, which would displace the drilling fluids, pushing them upwards, as well. In 2012, the decision was finally made to punch through to the lake, and once it had been breached, to pull the drill up, and let the lake water refreeze in the borehole. This was done at the end of the working (Antarctic summer) season, February, 2012. During the next working season (Nov. 2012 — Feb. 2013), they returned and pulled up a core of the frozen lake water that had been pushed up into the borehole. Everyone was excited, including the drillers, researchers, and Russian dignitaries. But, when they tested the ice meltwater back in a European lab, it was found to contain drilling fluids all the way to the center of the core, and the organisms found were all common contaminants in the drilling fluids. Unfortunately, the attempt to recover Lake Vostok water failed. And, to make it worse, the lake may have been contaminated, which might compromise any future study of the lake.

A group of researchers drilled into subglacial Lake Whillans, in West Antarctica (Figs. 1.2 and 8.5), at about the same time that the Vostok researchers were drilling into Lake Vostok, but with better results. In 2013, they used a hot water drill to rapidly drill through the 800 m of overlying ice, and then sent sterilized sampling containers down to collect water from the lake, and also collected sediments from the lakebed. Their metagenomic analysis yielded sequences from nearly 4,000 species of microbes. The primary ecosystem is based on the oxidation of ammonia, iron, sulfur, and methane, primarily by bacteria. Many are similar to those previously identified in surface lakes and sediments, and many are similar to those reported from Lake Vostok accretion ice, although there also are some major differences. However, the concentration of organisms reported from Lake Whillans was reported to be approximately 130,000 cells per milliliter, while in Lake Vostok, the concentration of organisms was much lower, from 10 to about 1,000 cells per milliliter (depending on the ice core section). Lake Whillans has a subglacial stream feeding it, as well as a subglacial outlet stream that leads to the nearby ocean, which lies under the Ross Ice Shelf. Therefore, it is possible that nutrients and microbes exchange between the bodies of water, which would explain the high concentrations of microbes within the lake samples. Also, Lake Whillans is under 800 m (2,625 ft) of ice, whereas Lake Vostok lies under more than 3,700 m (12,140 ft) of ice, which would generate very different pressures on organisms in the lakes. Additionally, Lake Vostok is completely below sea level, and is more than 1,160 m (3,820 ft) deep at its deepest point. Lake Whillans lies above sea level and is an average of 2 m (6.5 ft) deep. So, the habitats and ecosystems between the two lakes differ significantly. It is probable that each subglacial lake has a unique ecosystem.

FUTURE

Just a few decades ago, the ice on Kilimanjaro in Africa was more than 200 m thick. Today, it is almost gone. Glacier National Park, in the US will soon have no glaciers of the more than 100 that existed a century ago. Glaciers throughout the world are receding, and many have disappeared. In the Andes, Alps, and Himalayas, studies are underway on how to cope with the losses of the glaciers. Much of the human population relies on meltwater from glaciers for agriculture, drinking water, recreation, and hydroelectric power. The loss of these glaciers will impact on human, plant, and animal populations. In 1980, Antarctica was losing an average of 44 billion tons of ice annually. Since 2009, it has been losing an average of 278 billion tons annually, which already has caused measurable sea level rises worldwide. About 80% of the loss has been in West Antarctica, which is lower in elevation that East Antarctica. However, East Antarctica now shows signs of increased ice losses. The amount being lost in East Antarctica has been surprising to scientists, who had estimated far less loss from East Antarctica. It is very worrisome. The microbial releases from the melting are now beginning to be realized and studied. However, at current rates of loss of these resources and hazards, we may run out of time to study them before it is too late to determine their uses and dangers.

SOURCES AND ADDITIONAL READINGS

Augustin, L., C. Barbante, P.R. Barnes, et al. (55 authors). 2004. Eight glacial cycles from an Antarctic ice core. *Nature* 429: 623-628.

Castello, J.D., and S.O. Rogers. 2005. *Life in Ancient Ice*. Princeton, NJ: Princeton University Press.

Castello, J.D., S.O. Rogers, W.T. Starmer, C. Catranis, L. Ma, G. Bachand, Y. Zhao, and J.E. Smith. 1999. Detection of tomato mosaic tobamovirus RNA in ancient glacial ice. *Polar Biology* 22: 207-212.

Christner, B.C., J.C. Priscu, A.M. Achberger, C. Barbante, S.P. Carter, K. Christianson, A.B. Michaud, J.A. Mikucki, A.C. Mitchell, M.L. Skidmore, T.J. Vick-Majors, and the WISSARD Science Team. 2014. A microbial ecosystem beneath the West Antarctic ice sheet. *Nature* 512: 310-313.

Desmond, A., and J. Moore. 1991. Darwin, The Life of a Tormented Evolutionist. New York: W. W. Norton & Company.

D'Elia, T., R. Veerapaneni, and S.O. Rogers. 2008. Isolation of microbes from Lake Vostok accretion ice. *Appl. Environ. Microbiol.* 74: 4962-4965.

D'Elia, T, R. Veerapaneni, V. Theraisnathan, and S.O. Rogers. 2009. Isolation of fungi from Lake Vostok accretion ice. *Mycologia* 101: 751-763.

Knowlton, C., R. Veerapaneni, T. D'Elia, and S.O. Rogers. 2013. Microbial analysis of ancient ice core sections from Greenland and Antarctica. *Biology* 2: 206-232.

Ma, L., C. Catranis, W.T. Starmer, and S.O. Rogers. 1999. Revival and characterization of fungi from ancient polar ice. *Mycologist* 13:70-73.

Ma, L., S.O. Rogers, C. Catranis, and W.T. Starmer. 2000. Detection and characterization of ancient fungi entrapped in glacial ice. *Mycologia* 92: 286-295.

Petit, J.-R, J. Jouzel, D. Raynaud, N.I. Barkov, J.-M. Barnola, I. Basile, M. Bender, J. Chappellaz, M. Davis, G. Delaygue, M. Delmotte, V.M. Kotlyakov, M. Legrand, V.Y. Lipenkov, C. Lorius, L. Pépin, C. Ritz, E. Saltzman, and M. Stievnard. 1999. Climate and atmospheric history of the past 420,000 years from the Vostok ice core, Antarctica. *Nature* 399: 429-436.

Rogers, S.O., Y.M. Shtarkman, Z.A. Koçer, R. Edgar, R. Veerapaneni, and T. D'Elia. 2013. Ecology of subglacial Lake Vostok (Antarctica), based on metagenomic/meta-transcriptomic analyses of accretion ice. *Biology* 2:629-650.

Rogers, S.O., V. Theraisnathan, L.-J. Ma, Y. Zhao, G. Zhang, S.-G. Shin, J.D. Castello, and W.T. Starmer. 2004. Comparisons of protocols to decontaminate environmental ice samples for biological and molecular examinations. *Appl. Environ. Microbiol.* 70:2540-44.

Shtarkman Y.M., Z.A. Koçer, R. Edgar R, R.S. Veerapaneni, T. D'Elia, P.F. Morris, and S.O. Rogers. 2013. Subglacial Lake Vostok (Antarctica) accretion ice contains a diverse set of sequences from aquatic, marine and sediment-inhabiting Bacteria and Eukarya. PLoS ONE 8(7): e67221. doi:10.1371/journal.pone.0067221.

LINKS

https://en.wikipedia.org/wiki/European_Project_for_Ice_Coring_in_Antarctica
https://en.wikipedia.org/wiki/Ice_core
https://en.wikipedia.org/wiki/International_Geophysical_Year
https://en.wikipedia.org/wiki/WAIS_Divide
https://neem.dk/about_neem/
https://www.rt.com/news/226127-lake-vostok-russia-water/

12 Subglacial Lake Vostok

"I'm on a tightwire, one side's ice, the other's fire"

Leon Russel

WALKING ON THICK ICE

During the IGY, in 1957/58, one very remote station was set up by the USSR, close to Earth's geomagnetic pole, called Vostok Station (*Stántsiya Vostók;* translation, "Station East"), named after a warship with the same name. Vostok Station is approximately 1,400 km (900 mi) from the nearest coast and McMurdo Station, and approximately 1,200 km (800 mi) from the Amundsen-Scott South Pole Station (Fig. 12.1a). In addition to its remote location, it is situated 3,488 m (11,444 ft) above sea level, and is one of the coldest places on Earth. Temperatures in the summer (December/January) average from a high of -27 °C (-17°F) to a low of -38°C (-25°F), and in the winter (June/July) from a high of -61°C (-80°F) to a low of -70 (-95°F). The all-time high was -14°C (7°F), while the all-time low was -89°C (-129°F). During more than four months of the year, there is no sunlight at all, while during the summer months, the sun never sets. By the way, you can view the current weather on the internet by searching for Vostok Station weather (you can do the same for the Amundsen-Scott South Pole station, as well). It required a herculean effort for explorers to reach the site and set up the station. At that time, they could not fly into the site. They reached Antarctica by ship, and then they used large snowcats to drive to the site, which took several weeks. The snowcats were not only their transportation, but their lifeline, providing heat and electricity (from generators), while they pulled their living quarters, food, and fuel behind. At least two vehicles were always included on the journeys, in case one broke down along the way. This is a reminder of the dangers of working in the polar regions.

The initial purpose of Vostok Station was to record temperatures, snowfall, wind speeds, etc. It was essentially a weather station, as were many of the early stations. They also studied the snow and ice below them, as much as was possible at that time. From very early on they noticed the flatness of the area. The vast plain before them was more than 250 km (155 mi) in length, and more than 50 km (31 mi) wide, at its widest point (Fig. 12.1b). Even at that time, some (namely Andrey Kapitsa) hypothesized that a huge lake or lakebed might exist below the ice. He had used seismic methods to measure the thickness of the ice. From the soundings, it was clear the ice was several kilometers in thickness, and something other than rock existed at the bottom. There was no way to peer down through

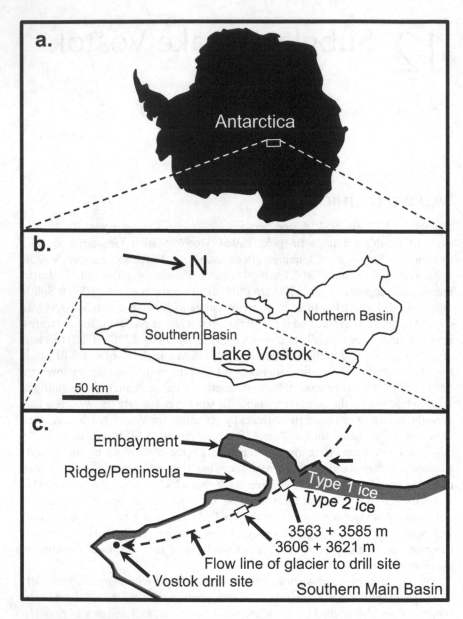

FIGURE 12.1   Maps of Antarctica (a) and Lake Vostok (b). The close-up map of Lake Vostok (c) shows the outlines of the southern portion of the lake (including the shallow bay, peninsula, and southern main basin), the direction of flow of the glacier, and the location of the ice coring site at Vostok Station. The extent of the type 1 and type 2 accretion ice is indicated.

the ice to determine what was below, so the presence of a lake remained a theory for many years. However, the thought of a lake below the ice was intriguing, and this set up the groundwork for further studies, primarily based on drilling ice cores. The first drilling began in 1970, ultimately reaching a depth of 952 m (3,123 ft). This core was just the beginning, and stimulated plans for deeper drilling. At about the same time, British researchers performed ground-penetrating radar studies of the area. A very sharp radar return was seen at the bottom of the ice, suggesting reflection off of a water surface. The idea of a lake under the ice was becoming a real possibility. Plans were made for deep drilling (see Fig. 8.5), which was begun in the 1980's, reaching 2,202 m (7,224 ft) in 1984 (3G ice core), 2,546 m (8,353 ft) in 1990 (4G ice core), and 3,623 m (11,886 ft) in 1996 (5G ice core). While the lake had not yet been reached, the ice and additional radar data indicated that they were within a few hundred meters of the lake surface (Figs. 8.5, 12.1c, and 12.2).

INITIAL CHARACTERIZATION OF THE VOSTOK ICE CORE

From the surface down to 3,310 m (10,860 ft), the ice was typical glacial ice, layer after layer having been laid down in chronological order (Fig. 12.2, 12.3, and 12.4). But, the ice changes below this level. For more than 200 m (656 ft), the ice is crushed, broken, and deformed. This is the basal ice, where the glacial ice was disturbed through interactions with the bedrock in the surrounding mountains and ridges. Then, at 3,538 m (11,608), the ice changes dramatically. It is more uniform and contains fine particulates not found in the basal ice above. This is

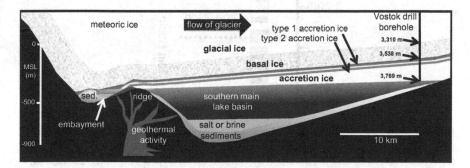

FIGURE 12.2 Cross section of Lake Vostok and the overlying ice. The glacial ice is from recent to approximately 1 million years old (at the bottom), including the lower 228 m (748 ft) of basal ice. The glacier moves at a rate such that the accretion ice builds up over a period of approximately 15,000 to 20,000 years. Therefore, at the position of the ice core site, the oldest accretion ice is closest to the bottom of the basal ice, while the youngest ice is immediately adjacent to the lake water. The type 1 and type 2 accretion ice layers, shallow bay, peninsula, southern main basin, possible sediment layers, possible marine layers, and possible hydrothermal activity are indicated. MSL = mean sea level.

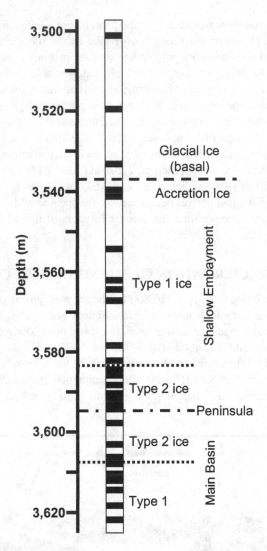

FIGURE 12.3 Details of the lower portion of the Vostok ice core, indicating the depths, basal ice, and accretion ice regions (type 1 ice, type 2 ice, bay, peninsula, and main basin). Ice core sections that have been studied by various research groups are noted as black bars within the ice core.

the beginning of the accretion ice that has formed from lake water frozen to the bottom of the glacier (Figs. 8.5, 12.2, 12.3). The accretion ice was studied by several research groups, including ours. The first 40–45 m (130–148 ft) of accretion ice contains high concentrations of several ions, including Na+, Cl-, and others, as well as fine mineral particles. This is called "type 1" accretion ice (Figs. 12.1c, 12.2, 12.3, and 12.4). It resembles coastal estuary water in its ion concentrations and suspended sediments. Evidence of hydrothermal activity also is present in

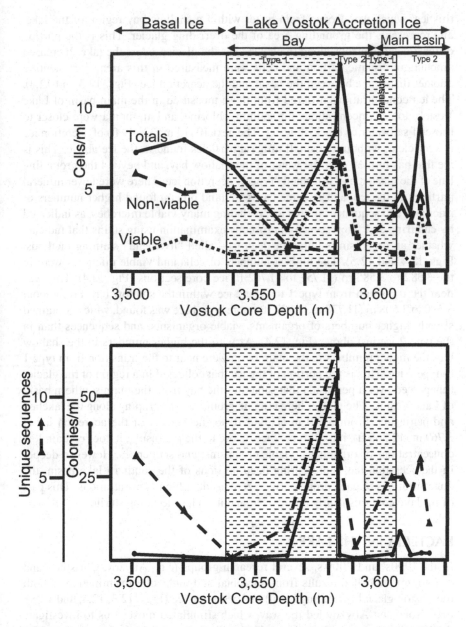

FIGURE 12.4 Graphs of our results from the Vostok ice core. The total numbers of cells, viable cells, and non-viable cells (as measured using a live/dead stain with fluorescence microscopy) are shown in the upper graph. The numbers of cultures (indicating viable organisms), and unique sequences (indicating species richness/diversity) are shown in the lower graph.

this ice, which was frozen from water within a shallow bay region of the lake, and represents the grounding area of the overriding glacier. This is the portion of the glacier that rubs along the lakebed as the glacier enters the lake. Increased amounts of organic carbon and cells were measured in this area, documenting another difference between the basal and the accretion ice (Figs. 12.3 and 12.4). The ion concentrations differed from those measured in the main basin of Lake Vostok. Ionic concentrations for sodium, chlorine, and ammonia were closer to those of saltwater or brackish water. The next 10–15 m (33–49 ft) of accretion ice is very clear, without the small mineral particles found in the ice above. This is ice that has formed over the water of the shallow bay, and beyond the grounding line of the glacier, and is termed "type 2" accretion ice. There were fewer mineral particles in the ice, and in some cases we (and others) found higher numbers of microbes and sequences in this ice, including many viable microbes, as indicated by culturing and fluorescence microscopy examination using stains that indicate whether an organism is living or dead (so-called "live/dead" staining methods; Figs. 12.4 and 12.5). The highest numbers of cells and viable microbes were in the 3,582 to 3,585 m (11,752 to 11,762 ft) ice core sections (Fig. 12.4). This was near the transition from type 1 to type 2 ice within the shallow bay. From about 3,590 to 3,608 m (11,778 to 11,837 ft), more type 1 ice was found, which contained slightly higher numbers of organisms, viable organisms, and sequences than in the type 2 ice just above (Fig. 12.4). As with the higher numbers in the shallow bay, the higher numbers in the main basin were near to the transition from type 1 to type 2 ice. This is the portion of the ice that collected in a region of the glacier that passed over a peninsula that separates the bay from the main southern basin of Lake Vostok. The glacier here is again grounded, or scraping along the lakebed and peninsula. From 3,610 m (11,844 ft) to the bottom of the accretion ice at 3,769 m (12,365 ft), it is all type 2 ice. Close to the peninsula, it does contain low concentrations of organisms, viable organisms, and sequences. However, deeper in the accretion ice, representing deep portions of the southern lake basin, the concentrations of each drop nearly to zero, as do all ion concentrations. This part of the lake contains nearly pure water, and it is close to being sterile.

BACTERIA IN THE LAKE

In the 1990s' and 2000's, several research groups (e.g., Abyzov, Christner, and our group) published results from microbial and molecular examination of both the Vostok glacial (including basal) and accretion ice (Figs. 12.3, 12.4, and 12.5). Professor Sabit Abyzov led the way, which stimulated most of us to investigate the Vostok ice core sections. He and his colleagues published micrographs of the bacteria that they found in the ice core sections. Most of their studies, as well as others, consistently reported that the glacial ice contained low concentrations of living organisms (e.g., fluorescence microscopy, culture assays, sequences). Our results were consistent with these reports, although we found that some of the accretion ice had much higher concentrations of these microbes (Figs. 12.4 and 12.5). These publications were the initial evidence that Lake Vostok was not

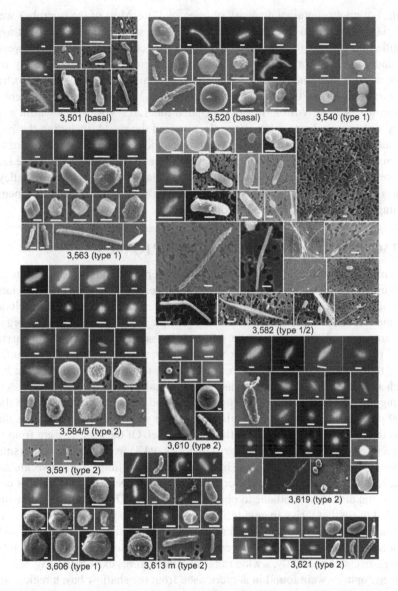

FIGURE 12.5 Collage of microbes that we observed from the Vostok ice core. The 3,501 m and 3,520 m core section is within the basal ice of the glacier, which is potentially more than one million years old. Cells here showed signs of damage and breakage, and most were dead, as indicated by culturing and staining methods. The other core sections are from the Lake Vostok accretion ice (lake water frozen to the bottom of the glacier). They range in age from 5,000 to 15,000 years old. The youngest sections are closest to the lake surface. The 3,582 m and 3,584/85 m core sections had the highest numbers and diversities of microbes for any of the sections. They also had the highest number of living organisms (as indicated by culturing and fluorescent staining methods). They were from a shallow bay in the lake, which has high levels of salts (i.e., ions) and nutrients. It also may contain hydrothermal vents.

sterile, but might contain a diversity of microbial life. Most of the microbes were bacteria, but there also were many fungi present, as well. The microbes that were identified are similar to those detected from other cold environments. However, it included some surprises, as well, including several bacteria that usually live in hot water environments. But, about the same time, geologists were reporting that earthquakes had been detected in Antarctica, including some that were in the vicinity of Lake Vostok. It was concluded that Lake Vostok lies within a rift valley system, similar to other rift valleys around the world, including the large rift valley in East Africa. Rift valleys are geologically active regions where the land masses are separating, such that in the middle of the region, part of the land subsides, forming valleys with steep sides. Often lakes and rivers fill the valleys, as they do in Africa. Lake Vostok appears to lie within a large rift that began forming tens of millions of years ago.

METAGENOMICS AND METATRANSCRIPTOMICS

Beginning in 2009, we decided to initiate a metagenomic and metatranscriptomic study using several sections of the Lake Vostok accretion ice. This yielded a large sequence data set gathered from the organisms present in the lake, and included information about some of the important biological processes that these organisms are accomplishing in the lake. Most of the biological action was occurring in the open water areas of the bay, which was consistent with previous studies. Many thermophilic organisms were found in the ice that accreted over the bay, which indicated that there also was hydrothermal activity in that region, supporting the conclusions of previous geological and biological studies. More than 3,000 unique sequences were obtained from the samples, including more than 1,600 that were used to identify to the species level. Of these, 95% were from ice representing the shallow bay. The large majority (94%) were bacteria, and a small percentage (6%) were eukaryotes, half of which were fungi. Only two archaea sequences were found, and both were marine methanogens (organisms that convert carbon dioxide to methane to obtain their energy). The mixture of organisms included those that thrive in cold temperatures (called psychrophiles and psycrotolerant organisms), those that prefer moderate temperatures (mesophiles), and a set of bacteria that only live in hot environments (thermophiles and thermotolerant organisms). Apparently, a wide range of temperatures exist in the bay. Most of the thermophiles were found in accretion ice from the shallow bay, a region with suspected hydrothermal activity based on other microbial and geological studies.

Bacteria that normally grow in other extreme environments: acidic (acidophiles), alkali (alkaliphiles), and salty (halophiles) environments also were detected. As with the bacteria that grew in hot regions, most often bacteria that prefer acidic, alkali, and salty regions, as well as those that were desiccation resistant, were found mainly in ice from the shallow bay. Within the ice that formed over the main basin, there were primarily bacteria that normally grow in cold freshwater, with only a few thermophilic and halophilic bacteria that may have floated over from the bay. Therefore, the main basin probably has little hydrothermal activity, and contains very cold, clear water; while the bay contains a source of hot, salty water, suggesting a hydrothermal vent system.

METABOLIC PROCESSES

Knowing what the organisms are doing in the lake is as important, or more so, than just knowing which ones are there. By comparing species characteristics, and comparing this to the gene sequences that we determined from the metatranscriptomic data, we deduced a great deal about the biological processes occurring in the lake. We could also deduce some components of the lake ecosystems that had not previously been measured. First, we looked for signs of autotrophs (organisms that can fix carbon from carbon dioxide, in order to make their organic molecules) and heterotrophs (organisms that cannot fix carbon from carbon dioxide, but instead require acquisition of organic molecules from other organisms). Bacterial species capable of carbon fixation from carbon dioxide exhibited one of two types of carbon fixation, the standard reductive pentose phosphate cycle (also present in the chloroplasts of plants, often called the Calvin cycle), and the reductive tricarboxylic acid (TCA) cycle (often called the Krebs cycle, also present in mitochondria, responsible for part of cellular respiration). Most bacteria (and mitochondria) have a TCA cycle (which is part of the system you, and most other organisms, use to make energy-containing molecules by combining oxygen with hydrogen ions and electrons), but they use the oxidative version, which gives off carbon dioxide (this is where the carbon dioxide comes from when you exhale). However, the reductive TCA cycle runs in the opposite direction and takes carbon dioxide and attaches it to other organic molecules, which is another method to fix carbon from carbon dioxide. Two other carbon-fixation pathways were identified from the metagenomic/metatranscriptomic data, the reductive acetyl-CoA pathway (the reversal of another common pathway) and the 3-hydroxylpropionic acid cycle, an ancient pathway used by only a few types of bacteria. The take home message is that there are several diverse biochemical systems in the organisms living in the lake. In other words, the lake is not sterile, and the ecosystem is complex.

Nitrogen, and its cycling in ecosystems, is essential for the production of proteins, nucleic acids, and other vital molecules in cells. Organisms and ecosystems must use a number of biochemical pathways and systems in order to acquire and utilize carbon and nitrogen. Nitrogen gas (N_2) in the atmosphere is nonreactive, and only certain bacteria are able to fix nitrogen (i.e., convert nitrogen gas to other compounds, primarily ammonium, NH_4^+). There were many species of nitrogen-fixing bacteria found in the accretion ice, including species of Actinobacteria, Cyanobacteria, Firmicutes, and Proteobacteria. The ammonium produced by this, and other, processes is used by many unicellular and multicellular organisms as a nitrogen source in a process called assimilation. As these organisms die, fungi and other organisms act on the nitrogen compounds to produce ammonium again, which reenters the nitrogen cycle. The ammonium is converted by nitrifying bacteria into nitrite ions (NO_2^-), and then into nitrate ions (NO_3^-) by other bacteria. Denitrifying bacteria perform the opposite reactions for these steps in the process of denitrification. Finally, nitrates can be converted back to nitrogen gas by some denitrifying bacteria, and a few types of bacteria (mainly a special group of bacteria, called planctomycetes) can take ammonium and nitrite to produce nitrogen gas, using a process called anammox. In addition to these main pathways and cycles, there were other processes indicated in the sequence data, including arsenite oxidation, arsenate reduction, iron oxidation, iron reduction, manganese oxidation, sulfur oxidation, and sulfur reduction. Each of these add to the likelihood of hydrothermal activity in the bay region. The bacteria identified by the metagenomic and metatranscriptomic data collectively form a viable, functional, subglacial aquatic ecosystem deep below the surface of the ice. Much of the energy in the system is probably coming from the hydrothermal activity. Some organisms are using this energy to fix carbon and nitrogen gases supplied by glacial meltwater as trapped atmospheric gases are being released as the ice melts. Viable organisms, as well as small amounts of organic carbon are being delivered to the lake by the melting glacial ice. From there, other organisms in the lake modify the carbon and nitrogen compounds, which are incorporated into amino acids, nucleotides, fatty acids, carbohydrates, and others, and ultimately used to synthesize proteins, DNA, RNA, phospholipids for membranes, sugars, polysaccharides, and other essential biomolecules needed for life.

EUKARYOTES IN LAKE VOSTOK

Eukaryotes, including multicellular species had been described from the accretion ice prior to the metagenomic and metatranscriptomic studies. Previously, we had cultured many species of fungi, mainly from accretion ice from the bay region. So, it was not surprising to find that half of the sequences from eukaryotes in the data were from fungal species. Fungi are major recyclers of carbon and nitrogen compounds in many ecosystems, and this is probably their major function in Lake Vostok. Also, some are pathogenic, parasitic, or mutualistic. One set of sequences was closest to a marine fungus found in a deep mid-ocean thermal vent. Several were identified that normally are associated with animals or plants. Many bacteria normally associated with specific animals and plants were present. Several of these had only previously been described as living only in the intestines of certain species of fish. Although sequences from fish were not detected in the data, fish have been discovered living in some subglacial bodies of water. There were other bacteria detected in the lake that are usually associated with sea squirts, clams, anemones, seaweeds, tubeworms, crustaceans, and other complex multicellular organisms. While these are indirect indicators of these organisms, they do advance the theory that complex multicellular eukaryotes live in Lake Vostok. More direct indications of complex eukaryotes were sequences from marine clams, sea anemones, tardigrades, rotifers, crustaceans, and aquatic arthropods. Additionally, several protist sequences were present, including those from the Excavata (closest to trypanosomes), Rhizaria (closest to a freshwater species), Alveolata (mainly phytoplankton), Stramenopiles (mainly phytoplankton), Heterokonta (mainly water molds), Cryptophyta (from freshwater and brackish water environments), and Amoebozoa (including *Naeglaeria*, free-living amoebae that can cause deadly, non-treatable infections in humans, wherein they destroy brain tissues). There also were sequences from plants. It is unknown whether some parasitic plants may exist in the lake, or whether these sequences represent pollen and plant fragments that were delivered to the lake by meltwater from the glacier.

LAKE WATER SAMPLING

As stated in an earlier chapter, the Russians halted drilling at Vostok from 1993 until the 2011–2012 season, primarily to go through the many different possibilities to sample the lake water, without contaminating the lake, and without contamination of the sample to be assayed. After many meetings, it was decided to go ahead in early 2012 to drill into the lake and then allow the lake water to rise in the borehole. It was hypothesized that the water in the borehole would remain clear of any drilling fluids. On February 5th, 2012, they drilled through the last few meters of ice and punched into the lake. Immediately, they pulled up the drill, packed up, and returned home, as it was the end of the working season in Antarctica. Then in the 2012–2013 season, they returned, sent the drill back down and retrieved a core from the frozen lake water that had risen up into the borehole the previous year. When the ice core sample was analyzed back in Europe, it contained drilling fluids and signs from the organisms that grow in the drilling fluids. They concluded that sampling of the lake water had failed. They had tested the drilling fluids previously, and had found at least 225 different bacteria present in the fluids. A plan was devised to perform deviation drilling around the end of the borehole to punch through the lake in another location. This was done in January of 2015, and a clean sample may have been obtained. However, no report of the findings was published, and funding for the project was discontinued. So, we will have to wait to see what happens with a future continuation of the study of Lake Vostok.

SOURCES AND ADDITIONAL READINGS

Castello, J.D., and S.O. Rogers. 2005. *Life in Ancient Ice*. Princeton, NJ: Princeton University Press.

Christner, B.C., G. Royston-Bishop, C.M. Foreman, B.R. Arnold, M. Tranter, K.A. Welch, W.B. Lyons, A.I. Tsapin, M. Studinger, and J.C. Priscu. 2006. Limnological conditions in subglacial Lake Vostok. *Limnol. Oceanogr.* 51: 2485-2501.

D'Elia, T., R. Veerapaneni, and S.O. Rogers. 2008. Isolation of microbes from Lake Vostok accretion ice. *Appl. Environ. Microbiol.* 74: 4962-4965.

D'Elia, T, R. Veerapaneni, V. Theraisnathan, and S.O. Rogers. 2009. Isolation of fungi from Lake Vostok accretion ice. *Mycologia* 101: 751-763.

Petit, J.-R, J. Jouzel, D. Raynaud, N.I. Barkov, J.-M. Barnola, I. Basile, M. Bender, J. Chappellaz, M. Davis, G. Delaygue, M. Delmotte, V.M. Kotlyakov, M. Legrand, V.Y. Lipenkov, C. Lorius, L. Pépin, C. Ritz, E. Saltzman, and M. Stievnard. 1999. Climate and atmospheric history of the past 420,000 years from the Vostok ice core, Antarctica. *Nature* 399: 429-436.

Rogers, S.O., Y.M. Shtarkman, Z.A. Koçer, R. Edgar, R. Veerapaneni, and T. D'Elia.
    2013. Ecology of subglacial Lake Vostok (Antarctica), based on metagenomic/meta
    transcriptomic analyses of accretion ice. *Biology* 2: 629-650.
Shtarkman Y.M., Z.A. Koçer, R. Edgar R, R.S. Veerapaneni, T. D'Elia, P.F. Morris, and
    S.O. Rogers. 2013. Subglacial Lake Vostok (Antarctica) accretion ice contains a
    diverse set of sequences from aquatic, marine and sediment-inhabiting Bacteria and
    Eukarya. PLoS ONE 8(7): e67221. doi:10.1371/journal.pone.0067221.

LINKS

https://en.wikipedia.org/wiki/Lake_Vostok
https://www.rt.com/news/226127-lake-vostok-russia-water/

13 Discoveries from Other Ice-Covered Lakes

"Some people look like frozen lakes; break the ice, there you will see a lively world!"

Mehmet Murat ildan

SUBGLACIAL ENVIRONMENTS

In 2006, satellite data showed that some regions of Antarctic ice would rise by several meters over a year or two. Then, it would sink over a period of months. Subsequently, it was discovered that there were subglacial lakes under these regions that increased in volume and pressure, which caused the overlaying ice to rise by several meters. This was followed by a release of the water, causing the level of the ice to drop. Furthermore, this phenomenon was cyclical. The first measurements of the amount of water involved were astounding. Over a period of 16 months, two subglacial lakes gained a total of 1.8 km³ (0.43 mi³) of water, equivalent to 1.8×10^{12} liters (4.7×10^{11} gallons). The water entered these lakes from subglacial streams. However, at first the outflowing streams were blocked, probably by ice and rock. Once the ice was pushed up and the pressure increased, this suddenly opened the blocked passages, and the water gushed through the opening until the ice dropped sufficiently to once again seal the outflow streams. These cycles repeat. During the past two decades, these subglacial environments have been the focus of many investigations. We now know that if the ice were to be removed from Antarctica, the land would look much like other parts of the Earth. There are mountains, canyons, valleys, streams, rivers, lakes, ponds, marshes, estuaries, islands, shallow seas, etc. The major difference is that ice is pressing on all of those areas, and this added pressure affects the land, as well as the water features. In fact, the weight of the ice in Antarctica and Greenland forces the land down, and it will take millennia for it to fully rise again once all of the ice has melted.

Nearly 400 subglacial lakes have been discovered in Antarctica, including Lake Vostok, which is the 16th largest lake on Earth by area (15,690 km² = 6,058 mi²). It is somewhat smaller in area than Lake Ontario (18,960 km² = 7,320 mi²), but contains more than three times as much water (1,640 versus 5,400 km³; 420 versus 1,296 mi³), making it the 6th largest lake by volume. Also, it is the fourth deepest lake on Earth (maximum depth 1,165 m, or 3,822 ft). The other Antarctic lakes range in size from small ponds up to those that are large, including

90°E Lake, Lake Sovetskaya, and a third large lake, yet to be named, which are 2,000, 1,600, and 1,250 km^2 (772, 618, and 483 mi^2) in area, respectively, and may be as deep as Lake Vostok. All are within approximately 1,000 km (620 mi) of Lake Vostok. They also are in the vicinity of a large subglacial canyon system that is larger than the Grand Canyon. Other smaller lakes that have been named are Lake Untersee (covering an area of 11.4 km^2 = 4.4 mi^2) and Lake Hodgson (area of 3 km^2 = 1.2 mi^2).

The organisms that live in these subglacial environments have become adapted to extreme conditions, at least compared to those living in temperate regions at the surface. No light penetrates the ice. The temperatures are usually below 0°C (32°F), although the temperatures at the bases of the ice formations are often close to 0°C (32°F), while temperatures at the surface of the ice can range between well below freezing to well above. Often, ion levels in subglacial lakes are in the range of brackish water, although the main basin of Lake Vostok has ion concentrations approaching that of pure water. One might think that this would be good for the organisms, but it presents many challenges for them, including the import and export of vital chemicals, and the maintenance of cell integrity due to differences in the osmotic pressures of such water. Some of the lakes, including Lake Vostok, have high concentrations of nitrogen and oxygen, up to 50 times that found in surface lakes. This causes rapid oxidation of the biochemicals in the cells, and inhibits growth of anaerobic organisms. This is another thing that affects cell viability. The lakes also are under tremendous pressure, which in Lake Vostok is estimated to be approximately 350 atmospheres (5,143 pounds per square inch; 3,616 kilograms per square meter). This would crush a human. We tested some of the organisms cultured from glacial and Vostok accretion ice at several different temperatures and pressures. All of them grew down to 4 to 8°C, and could withstand pressures equal to those in Lake Vostok. However, they grew faster at temperatures of 15 to 22°C and pressures closer to 1 atmosphere. Therefore, these were not psychrophiles (organisms that only grow at low temperatures) or barophiles (organisms that grow only under high pressures). They were simply highly adaptable organisms.

During the 2012–2013 Antarctic research season, an attempt was made to obtain a water sample from Lake Ellsworth (see Fig. 1.2 for location). It is smaller in area than Lake Vostok (covering 30 km^2 = 11.5 mi^2) in area, and in depth (maximum depth 150 m = 505 ft). It lies under 3.4 km (2.1 mi) of glacial ice, so it is almost as deep as Lake Vostok, but its surface is more than 1,400 m (4,600 ft) below sea level. Remember that the surface of Lake Vostok is 200 m (656 ft) below sea level. Lake Ellsworth and Lake Vostok are both under high pressure, and lie below sea level, which could lead to intrusion of saltwater into the lakes, and there are indications of this in Lake Vostok. Both lakes also are under high pressure because of the mass of ice on top of them, and they have very deep basins. When an attempt was made to sample from Lake Ellsworth in the 2012–2013 season, things did not go according to plan. After several days of attempts to solve some problems in the drilling, the plan to sample from the lake had to be abandoned. Future attempts are likely, but these are multi-million-dollar projects, so there will be intense scrutiny on future attempts to drill into Lake Ellsworth.

During the same season, researchers drilled into subglacial Lake Whillans using methods to assure clean samples were obtained. It is smaller than Lake Vostok, covering an area of 60 km² (23 mi³), with an average depth of 2 m (6.5 ft). The surface of Lake Whillans is covered by 800 m (2,625 ft) of ice, and is above sea level. Therefore, it is a very different than Lake Vostok. These two subglacial lakes are exposed to a number of different conditions, and as expected, the communities of organisms present in each lake were significantly different (Fig. 13.1). While species of proteobacteria predominate in both lakes, and Actinobacteria and Bacteroidetes also were in moderate numbers, Firmicutes were much higher in Lake Vostok, and Archaea were much higher in Lake Whillans, indicating the uniqueness of each lake.

The approach to them differed. To assure that they could reach Lake Whillans within the short Antarctic drilling season, they used a hot water drill to quickly drill through the 800 m (2,625 ft) of glacial ice over a four-day period. To minimize contaminating organisms, the drill water was constantly recirculated through a filtering system, UV-irradiated, and heated to 90°C. In addition, all drilling equipment and instruments were treated with a hydrogen peroxide solution. As the drill came close to the last of the ice above the lake, they slowed the drilling, and monitored the level of water in the borehole. When the level suddenly rose, they knew that the lake water had pushed up in the borehole due to the pressure in the lake. Once the lake was breached, they cautiously raised the drill head, waited for the pressure to stabilize, and then sent down sterilized sampling devices to collect water and sediment from the lake.

The concentration of microbes in Lake Whillans was reported to be 130,000 cells per ml, which was much higher than the microbial concentrations in Lake Vostok, which ranged from just a few cells to approximately 1,000 cells per ml. The ionic concentrations for sodium, chlorine, and ammonia in Lake Whillans were close to those of saltwater or brackish water. More than 3,900 unique bacterial and archaeal sequences were detected (Fig. 13.1). This included a large group of beta-proteobacterial species (as well as other types of proteobacteria), Bacterioidetes, Actinobacteria, and smaller numbers in a few other bacterial phyla. The taxonomic diversity indicates a mixture of freshwater, sediment-dwelling, and marine microbes, some of which are usually associated with fish and other eukaryotes. The Actinobacteria are a widespread group of bacteria found commonly in soil. They have been found in Antarctica by many researchers, as well as the Greenland ice sheet by us. Although Actinobacteria make up a small proportion of the bacteria isolated from ice, they have been found in some of the oldest ice layers, and also in the ice of surface and subglacial lakes. Actinobacteria may be capable of surviving for thousands of years in the ice or perhaps even longer. Members of this group of bacteria produce valuable antibiotics (including streptomycin, aureomycin, terramycin, and chloromycetin). Some of these from ice might produce antibiotics currently unknown to man. Perhaps a potent anticancer drug may be a byproduct of such an organism, or others that are being isolated from ice cores.

Stream systems connect Lake Whillans to other subglacial lakes upstream, and downstream to the ocean. Additionally, the ice above Lake Whillans rises

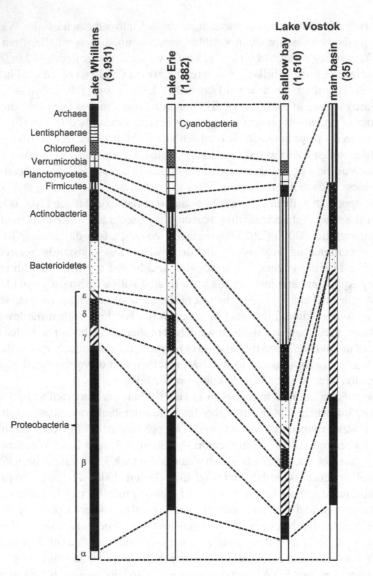

FIGURE 13.1 Proportions of unique (indicative individual species) bacterial (and archaeal in Lake Whillans) species (based on sequences) for three lakes: subglacial Lake Whillans, Lake Erie, and two regions (shallow bay and main basin) of subglacial Lake Vostok. Betaproteobacteria (β), Bacterioidetes, and Actinobacteria predominate in Lake Whillans, while Alphaproteobacteria (α), Betaprotobacteria (β), and Gammaproteobacteria (γ) are most frequent in Lake Erie (freshwater) and Lake Vostok main basin (freshwater). Firmicutes and Actinobacteria predominate in the shallow bay in Lake Vostok. Cyanobacteria in Lake Erie and in the shallow bay also are in high proportions, although the species differ. Numbers in parentheses indicate the total number of unique sequences (indicating species) in each sample. A small number of Deltaproteobaceteria (δ) and Epsilonproteobacteria (ε) also were present in Lake Whillans, Lake Erie, and the shallow bay of Lake Vostok, but were completely absent from the main basin of Lake Vostok.

and subsequently falls by as much as 5 m (16.5 ft), which indicates filling of the plugged lake, followed by unplugging and rapid release of water from the lake occurs periodically. Therefore, there is influx and efflux of water through the lake, leading to import and export of nutrients and organisms. Another research group drilled a hole through the Ross Ice Shelf into the underlying ocean, close to the grounding line of the glacier, which was near the outlet stream from Lake Whillans. In addition to the diversity of microbes that they observed microscopically, they sent a video camera down the borehole, and observed crustaceans (shrimp), jellyfish, and fish, some of which were up to 20 cm (nearly 8 inches) in length. This was a surprise. Some of the microbial life from this region may be able to reach Lake Whillans, or, alternatively, some of the microbial life from Lake Whillans, probably flows into this region. Whether larger organisms can move freely among the subglacial lakes is unknown. However, at this time, the results from Lake Whillans and the video from below the Ross Ice Shelf leave no doubt that there is a diversity of life, including complex multicellular organisms, under the ice in Antarctica. Subglacial lakes have also been found in the Arctic, so analogous systems also probably exist there.

SURFACE LAKE ICE

Worldwide, a large number of lakes freeze during the winter, and some lakes are frozen continuously for several years at a time. This is dependent on latitude and altitude, as well as with ion concentrations and lake size and depth. We studied ice from several lakes, including Lake Erie (the only Great Lake that freezes over completely during some winters), as well as several Antarctic surface lakes. As we were performing the metagenomic and metatranscriptomic study of Lake Vostok, we decided to perform the same procedures on a sample of surface ice from Lake Erie. This also became a test of our techniques, because we subjected the Lake Vostok and Lake Erie samples through the same processes at the same time, and they were sent for commercial NGS processing at exactly the same time to the same facility. It was clear from the sequencing results that the two lakes were very different, as expected (Figs. 13.1 and 13.2). The sequences indicated that the shallow bay in Lake Vostok contained a mixture of microbes, some of which grew in hot water, some in cold water, some inhabited sediments, some grew in salty water, others under acidic conditions, and some in alkali conditions; but not one of them was the result of human activity. The sequences from the Lake Erie ice sample showed a wide range of aquatic bacteria, including many diverse cyanobacteria; as well as eukaryotes, including diatoms, and other aquatic protists. The presence of diatoms and cyanobacteria had been previously well characterized in microbiological and ecological research throughout more than a century of research of Lake Erie. In fact, large cyanobacterial blooms (some of them toxic) have been well documented in the lake. The Lake Erie ice also had a large proportion of organisms associated with human habitation in and around the lake. There were many sequences from agricultural plants and animals (e.g., pigs, cows, and chickens, cereal grains, grasses, and legumes). Corn, wheat, and soybeans are

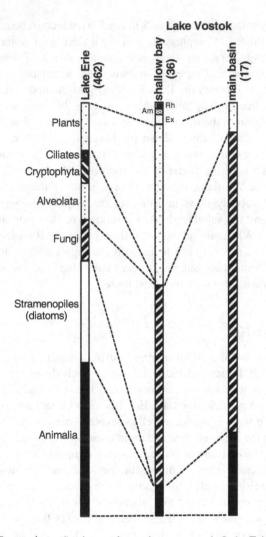

FIGURE 13.2 Proportions of unique eukaryotic sequences in Lake Erie and two regions of Lake Vostok (Lake Whillans was not evaluated for eukaryotes). Animals (including rotifers and other microscopic species) and diatoms predominant in Lake Erie. While small animals (e.g., tardigrades, rotifers, anemones) also are present in Lake Vostok, fungi (primarily ascomycetes) are present in the highest proportions. Numbers in parentheses indicate the total number of unique sequences (i.e., species) in each sample. Am = Amoebae; Ex = Excavata; Rh = Rhizaria.

major crops in this region. Also present were sequences frequently found in the guts of animals, including humans. There also were sequences from pathogenic organisms, indicative of human and animal waste. Sequences from a few thermophiles also were found, which was surprising at first. However, these may have come from the cooling water from industry and power plants along the lake shore.

Surface lakes also exist in Antarctica. These are lakes periodically open to the environment when some of the ice melts. All of these are within short distances of the coast, and at low elevations. We initiated a study of the ice from surface lakes in Antarctica. Ice from eight surface lakes around Antarctica were sampled by Professor David Gilichinsky and colleagues. We subjected the samples to metagenomic and metatranscriptomic analysis, and found a great deal of variation in the ice. The lake closest to the ocean and a research station was located at Bellinghausen, on an island of the Antarctic peninsula. Sea spray and foam sometimes reach the lake; and it was the lake with the largest diversity and concentration of microbe sequences. The ice from the other lakes had much lower concentrations and diversity of microbes. The lowest were from samples in the lake furthest from the ocean and human habitation (present or past). The microbes in these lakes have not yet been extensively evaluated, but they undergo freeze-thaw cycles frequently, which is physiologically stressful. However, being at temperatures above freezing for part of the year, and growing at one atmosphere of pressure, they are under less pressure than the organisms in the subglacial lakes. Therefore, the expectation is that the communities will be different than those in the subglacial lakes. Initial indications are that they are very different.

SOURCES AND ADDITIONAL READING

Christner, B.C., J.C. Priscu, A.M. Achberger, C. Barbante, S.P. Carter, K. Christianson, A.B. Michaud, J.A. Mikucki, A.C. Mitchell, M.L. Skidmore, T.J. Vick-Majors, and the WISSARD Science Team. 2014. A microbial ecosystem beneath the West Antarctic ice sheet. *Nature* 512: 310-313.

Christner, B.C., G. Royston-Bishop, C.M. Foreman, B.R. Arnold, M. Tranter, K.A. Welch, W.B. Lyons, A.I. Tsapin, M. Studinger, and J.C. Priscu. 2006. Limnological conditions in subglacial Lake Vostok. *Limnol. Oceanogr.* 51: 2485-2501.

Rogers, S.O., Y.M. Shtarkman, Z.A. Koçer, R. Edgar, R. Veerapaneni, and T. D'Elia. 2013. Ecology of subglacial Lake Vostok (Antarctica), based on metagenomic/meta transcriptomic analyses of accretion ice. *Biology* 2: 629-650.

Shtarkman Y.M., Z.A. Koçer, R. Edgar R, R.S. Veerapaneni, T. D'Elia, P.F. Morris, and S.O. Rogers. 2013. Subglacial Lake Vostok (Antarctica) accretion ice contains a diverse set of sequences from aquatic, marine and sediment-inhabiting Bacteria and Eukarya. PLoS ONE 8(7): e67221. doi:10.1371/journal.pone.0067221.

Siegert, M.J., R. Kwok, C. Mayer, and B. Hubbard. 2000. Water exchange between the subglacial Lake Vostok and the overlying ice sheet. *Nature,* 2000. 403: 643-646.

Stewart S.R., N. Ross, J.S. Greenbaum, D.A. Young, A.R.A. Aitken, J.L. Roberts, D.D. Blankenship, S. Bo, and M J. Siegert. 2016. An extensive subglacial lake and canyon system in Princess Elizabeth Land, East Antarctica. *Geology* 44: 87-90.

LINKS

https://en.wikipedia.org/wiki/Lake_Ellsworth_(Antarctica)
https://www.scientificamerican.com/article/discovery-fish-live-beneath-antarctica/

14 Life Is Everywhere

"In three words I can sum up everything I've learned about life. It goes on"

Robert Frost

HARD TIMES

Humans and their agricultural plants and animals are rather fragile organisms. They require shelter, tending, protection, and constant nourishment. If the weather changes suddenly, many may be injured, weakened, become infected, and die. Here is just one example. Beginning in the year 535 CE, the weather in the Northern Hemisphere changed dramatically. It still holds the record for being the coldest period in the past 2,000 years. Dust was present in the atmosphere, there was a thick haze in the sky, the sun dimmed considerably, and tree rings correlated to the time showed very slow or no growth, crops failed, and there were snowfall events, even in the summer months. The problems continued for many years, and temperatures only returned to normal after 550. There were widespread food shortages; leading to famine, disease, and death. A major plague epidemic struck Europe. Some cities and civilizations, including that of the Maya, went into decline around this time. Recent ice core analyses indicate that there were high sulfate concentrations in the atmosphere. The hazy skies and sulfates in the atmosphere point to a huge volcanic eruption in 535, followed by another one (probably from a different volcano) in 539. The crop failures, starvation, diseases, and deaths of plants and animals, including humans, indicates that rapid changes in the environment can be devastating to species that were unable to rapidly adapt to those changes. The species and individuals that did survive were the adaptable ones. Many of these were probably able to survive at this time by changing their metabolic activities. During the coldest times, some may have gone dormant.

A more catastrophic event provides another example of this. A huge asteroid impact 66 million years ago led to the extinction of all dinosaur species, as well as about 75% of all species on Earth. That extreme stress was too much for species that could not adapt to the rapid environmental changes that the impact caused, including the initial blast, shock waves, flaming rocks raining down, earthquakes, fires, smoke, acid rain, lack of sunlight, temperature fluctuations, vulcanism, and other phenomena. But, a large number of species did survive, including some small mammals (which eventually led to us), birds (close relatives of dinosaurs), reptiles, amphibians, fish, and many types of microbes. In fact, the fungal and bacterial populations expanded after that event, likely due to all of the biological matter made available through the deaths of plants, animals, and other microbes worldwide. Again, the organisms that survived to reproduce were the most adaptable ones.

ADAPTABLE SPECIES IN ICE

The biological inclusions that we and other researchers have been finding in ice around the world can be put into two categories: dead organic materials deposited onto the ice by wind and/or precipitation; and living individuals of species that can rapidly adapt to changes in their environments. In fact, these species can adapt to many different environments (Table 14.1). We found sequences from tardigrades in the shallow bay of Lake Vostok. Tardigrades (commonly called "water bears") are tiny (about 0.5 mm in length; 0.02 inches) segmented animals with eight legs, composed of about 40,000 cells. They are extremely adaptable. Some of them

TABLE 14.1

Characteristics of Organisms Found in Environmental Ice (Glacial, Subglacial, and Surface Lake)

Growth Parameter		Glacial Ice	Subglacial Lake	Surface Lake
	Thermo	+ (v)	+ (h)	+ (i)
Temperature	Meso	+++	+++	+++
	Psychro	+++	+++	+++
	High	+++	+++	n/d
Pressure	Moderate	+++	+++	n/d
	Low	n/d	n/d	n/d
	Saltwater	+	++	n/d
Salinity	Brackish	+	++	n/d
	Freshwater	+++	++	n/d
	Alkalai	n/d	+	n/d
pH	Neutral	n/d	+++	n/d
	Acid	n/d	+	n/d
Freeze/Thaw Resistant		++	+++	n/d
Desiccation Resistant		n/d	+	n/d
Soil/Sediment Inhabiting		++	++	n/d
Most Prevalent Bacteria		Proteobacteria, Firmicutes	Proteobacteria, Actinobacteria	Proteobacteria, Cyanobacteria
Most Prevalent Archaea		Methanogens	Methanogens	Methanogens
Most Prevalent Eukaryotes		Fungi	Fungi	Animals, Diatoms

Note: One plus, + = a few organisms were present; ++ = a moderate number of organisms were found; +++ = a large number of organisms were found; n/d = not determined. Possible sources of thermophiles: v = volcanoes; h = hydrothermal activity; i = industrial plants can survive temperatures almost to absolute zero (-273°C, -460°F), and others can survive up to 150°C (300°F). In addition to being resistant to extremes of temperature, they survive desiccation, starvation, high pressures, low pressures, lack of air, radiation, and other extreme conditions. Some have survived exposure to outer space. We found other complex (although small) multicellular organisms in the Lake Vostok accretion ice as well, including sea anemones (hydra), crustaceans (small shrimp), mollusks (small clams), and rotifers. Rotifers are small animals that can survive desiccation, cold temperatures, starvation, radiation, and high pressures. Hydras, crustaceans, and mollusks also have unique adaptive traits. There were also indications of fish, because sequences from organisms associated with fish intestines were found in the accretion ice. Many cold-water species of fish have been described from polar regions, including some that live beneath ice shelves. Some fish, crustaceans, and other organisms have been found living in the Marianas trench, at depths of more than 7,000 m (23,000 ft).

Many of the microbes found in ice also are extremely hardy. In 2005, a member of the bacterial species *Carnobacterium pleistocenium* was isolated from 32,000-year-old Alaskan pond ice. Other species of *Carnobacterium* have been found in cave ice, glacial ice, and subglacial lake samples that were from a few hundred to over 100,000 years old. Species in this bacterial genus are frequently found in polar regions, and survive high pressures, as well as multiple freeze-thaw cycles. They have been found living as deep as 2,500 m (8,200 ft) in the ocean, where the pressure is approximately 240 atmospheres. The pressure in Lake Vostok, where *Carnobacterium mobile* was detected, is at 350 atmospheres. Some species in this genus grow on animals, but most are not pathogenic. The species that was found in Lake Vostok accretion ice is usually associated with shrimp or fish. Sequences from crustaceans were found in Lake Vostok accretion ice, as well as sequences from several species of bacteria that are usually found in fish intestines. Other species within this genus also are associated with fish.

While most research has concentrated on bacteria that grow in extreme conditions, many eukaryotes, including many species of fungi, also are found in extreme environments. We have grown both bacteria and fungi from ice, including from some ice that is well over 500,000 years old. One sequence that we found in the accretion ice from Lake Vostok was very closely aligned with a sequence from a fungus that was described as growing near a hydrothermal vent in a deep ocean site. Lake Vostok appears to have hydrothermal activity, so this fungus may be associated with those areas in the lake. Being frozen in ice, and then being thawed at a later time, takes a great number of adaptations to survive. Many extremophiles exhibit metabolic activity even when frozen, including some that exhibit metabolic functions down to -80°C. Most produce tough cell walls or spores, and have membrane properties that protect the cells from breaking open during freezing and thawing events. Some are known to have antifreeze properties, including synthesis of proteins and/or polysaccharides that limit the growth of ice crystals, such that only small crystals are formed. Others excrete compounds that produce a liquid layer around the cells. Usually, the organisms are unicellular, or if they are multicellular, they are very small, although some arthropods, amphibians, and reptiles are known to survive freezing and thawing. However, the number of unicellular microbes that can survive long-term freezing, followed by thawing is much larger than for any other types of organisms. Some have survived more than a million years encased in ice. In 2007, a research group isolated a living bacterium from 8 million-year-old ice from the Beacon and Mullins valleys of Antarctica, and another group isolated a living bacterium from 3.5 million-year-old permafrost.

SENSITIVITY TO FREEZING, THAWING, AND TEMPERATURE

We tested some of the bacterial and fungal isolates that we had isolated from various Greenland and Antarctic ice core sections for their growth at different temperatures. They all were resistant to freezing and thawing, and a few even required occasional freeze-thaw cycles to start growing again, after they had become dormant. We tested their growth at several temperatures (4°C, 8°C, 15°C, 22°C, and 37°C). Most grew at every temperature, except 37°C, which killed almost all of them. This indicated that they needed cooler temperatures, and that they were probably not human pathogens, which usually grow well at 37°C. Most grew slowly at 4°C, but they all survived at that temperature. Most grew a bit faster at 8°C, and even faster at 15°C, but some of those grew slower at 22°C. This indicated that the majority of bacteria and fungi that were isolated from the ice were psychrotolerant, that is, they grew at low temperatures. A few grew only at 4°C and 8°C, but failed to grow at 15°C and above. These were true psychrophiles, defined as organisms that can only grow below 15°C.

One of the reasons that we probably isolated more psychrotolerant organisms than psychrophiles is that the original culturing was primarily performed at 15°C. Some cultures were started at 4°C and 8°C, but any microbes that grew, did so very slowly (over periods of many months). For our graduate students working on these projects, it made them anxious when they examined their culture plates week after week, and saw nothing (or very little) growing. So, out of necessity, and after a few experiments, we decided to use 15°C as the starting point for cultures. More things grew and we still were able to recover psychrotolerant organisms, as well as some psychrophilic organisms. The graduate students started smiling more often, and, more importantly for them, they completed their projects, and were awarded their degrees.

Sabit Abyzov and his coworkers have been isolating bacteria from ice and permafrost for several decades, longer than anyone else. They isolated viable bacteria, as well as filamentous fungi and yeasts, from the Antarctic ice sheet. *Pseudomonas* species (common soil bacteria that do not form endospores) were most often isolated from younger ice layers, whereas a variety of endospore-forming bacteria were recovered from deeper (older) layers. Up to 7% of the microbial biomass in the upper 100 m (238 ft) of the ice were members of the genus *Pseudomonas*. Others also have isolated endospore-forming bacteria, non-spore-forming bacteria, and Actinobacteria from Antarctic ice. Different species of endospore formers occur throughout the core. Because of the stability of their spores, it should not be a surprise that endospore-forming bacteria predominate in the oldest ice horizons. Strangely enough, many spore-forming bacterial isolates from ice grow abundantly at 37–40°C, which is atypical for many modern spore formers. Many of the spore-forming species were isolated from ice approximately 200,000 ybp. The deeper and thus the older the core, the greater the variety of spore formers. This date corresponds to a global warming period on Earth, that extended from about 200,000 to 250,000 years ago, which preceded the Eemian warm period that occurred between about 105,000 and 120,000 years ago, as well as our current warming period.

PRESSURE TESTS

We also tested some of the organisms for their ability to grow under high pressure conditions. After all, they were exposed to great pressures in the glaciers, as well as in the water of Lake Vostok, and other subglacial lakes, rivers, and streams. So, it was assumed that they would be able to grow at high pressures. The cultures were placed inside pressurized vessels (called pressure bombs) at several different higher pressures, and in every case, they grew. However, all of them grew faster at normal atmospheric pressure. This indicated that they tolerated high pressures (i.e., they were barotolerant), but they did not require high pressures, as do baro-philes, normally found in deep oceans or deep underground. Again, this outcome could have been caused by the fact that normally when we cultured the organisms in our lab, we did so at normal atmospheric pressure. Nonetheless, all of this leads to the same conclusion, that these organisms can survive, and even thrive, in a broad range of conditions. In other words, they are very adaptable, and this is a notable characteristic of the microbes that have been isolated and sequenced from ice cores.

OTHER ADAPTATIONS

Microbes similar to those found in temperate and tropical regions have been isolated from polar ice cores. This implies that they can survive exposure to a broad range of temperatures, again pointing to the most common characteristic of organisms isolated from polar ice sources, they all are very adaptable species. Also, perhaps surprisingly, photosynthetic bacteria have been found in deep (and dark) frozen permafrost. One wonders why photosynthetic organisms would be found living deep into the dark permafrost. It might be that the genes/proteins normally associated with photosynthetic processes are functional in other vital processes. Certainly, this combination of conditions and microbes is leading to some interesting and intense evolutionary processes. For example, several non-photosynthetic organisms have been found with chloroplasts and/or chloroplast genes. For example, trypanosomes, which cause parasitic diseases in humans (and other animals) have genes from chloroplasts in their genomes. Some of those genes are still functional! The reason for this is that their ancestors were produced by an endosymbiotic event between a non-photosynthetic protist cell and a green algal cell (both eukaryotic cells). The same event also produced the ancestor of the photosynthetic organism, *Euglena*. Over millions of years, trypanosomes became parasitic, and no longer needed to utilize their photosynthetic capabilities, because they were obtaining their organic compounds from their hosts, and the photosynthesis genes slowly mutated until most lost their functions and disappeared. Why chloroplasts and some of their genes were retained in some of these parasites is mostly a mystery, but their retention indicates some sort of selective advantage for such organisms. This leads to the conclusion that the diversity of microbes in nature far surpasses what we know presently. Therefore, perhaps we should not be surprised by these discoveries.

ANABIOSIS AND SPORES

The Dutch naturalist Anton van Leeuwenhoek (1632–1723) was a very famous scientist, and indeed probably the first true microbiologist. Although his ground-glass lenses are primitive by today's standards, they made it possible for him to draw and describe microbes (which he termed "animalcules") for the first time. He also was the first person to investigate long-term storage of microorganisms. By keeping the microbes absolutely dry, they could be revived after several months of storage, during which time they showed absolutely no signs of life. In 1702, he speculated about the possibility of what he called "latent life," which has since been called cryptobiosis, anabiosis, or stasis. Such conditions exist when an organism shows no visible sign of life, when its metabolic activity is very low, or comes to a reversible standstill. Since that time, much has been written about the revival of ancient life, while attempting to answer the question of just how long life can survive in an anabiotic state. The morphologies of cells in this state often differ from those in the same species under active growth conditions. Therefore, some microbes appear to have genetic switches that send them into a state of anabiosis under specific conditions. Many fungi can produce more than one type of spore, and can produce more than one cellular morphology, depending on temperature, nutrient level, and other factors. The process of freezing in ice apparently is one of the conditions whereby some microbes make the switch to an anabiotic, or an anabiosis-like, state.

Many of the bacteria detected in ancient substrates including ice produce endospores, which are extremely durable resting structures. They can survive for many thousands of years. Viable endospores of *Bacillus* species were present in soil particles adhering to the roots of 320-year-old dried plant specimens in the Herbarium of the Royal Botanic Garden, in London. Based on calculations at the time, it was estimated that soil would still contain viable spores even after 1,000 years, and that a small proportion of the spores might be capable of surviving to 10,000 years. Viable thermophilic bacteria have been recovered from ocean basin sediment cores 5,800 years old. Approximately 25–75 bacteria per gram of sediment were found. Viable endospores of another group of bacteria (thermoactinomycetes) were recovered at a concentration of 10,000 spores per gram from a 2,000-year-old archaeological site. It has been deduced that endospores covered by just a few centimeters of soil would be protected from ultraviolet radiation, and if stored at temperatures approaching absolute zero (-273°C) might be able to survive for millions of years. These conditions are similar to those found in comets and on several of the planets and moons in the Solar System. It has been speculated that endospores of *Bacillus subtilis*, even if exposed to the high vacuum and low temperature of interstellar space, could survive for millions of years if coated with a thin film of refractive material to protect them from radiation.

Experiments have been conducted to determine the half-life of bacterial endospores by determining just how much total gene damage would accumulate in endospores of *B. subtilis* exposed to background radiation. No significant difference in damage between fresh spores and those stored for 16 years was found. Based on these results, a half-life of 7,000 years was calculated for *B. subtilis* endospores. If one assumes an exponential death rate, a large population of viable endospores would still be detectable after several hundred thousand and possibly for millions of years. The reports of the isolation, growth, and DNA characterization of bacteria from 25- to 40-million-year-old amber and from 250-million-year-old salt crystals, if true, push the theoretical limit to hundreds-of-millions of years for some species entrapped in specific matrices, including ice. All of these expectations assume that the immediate environment protects the cells from ultraviolet irradiation, oxidation, and chemical damage common with hydrated nucleic acids and other biological molecules. Such protected conditions exist in glacial and sheet ice around the world. Viable and dead microorganisms including fungi, yeasts, protists, algae, bacteria, and viruses, have been detected in ice up to eight million years old. If one assumes that many of these reports are not due to contamination by recent microorganisms, then the organisms are able to protect or repair their DNA in ways that currently are not fully understood.

In general, the probability of recovering living microorganisms decreases with depth in the ice. However, scanning electron microscopy (SEM) reveals the presence of bacterial cells in meltwaters in which no viable isolates were recovered, clearly demonstrating that nonliving bacteria also are present in ice. Other methods (fluorescent methods utilizing selective staining) for detection of viable life forms are much more sensitive than direct culturing, and reveal that living cells are present even in the oldest ice examined to date (millions of years old). Of course, viability of the cells is dependent on a number of factors, including how they were deposited in the glacier, what their original condition was at the time of deposition, the temperature consistency, whether the ice warmed and melted, and others.

YEASTS AND FILAMENTOUS FUNGI

Bacteria and archaea are not the only microorganisms capable of prolonged survival in ice. Yeasts and filamentous fungi also are isolated frequently from ice and permafrost, some of which date back to the mid- to late-Pliocene Epoch (1.8–3.0 million years old). The yeasts isolated all are similar to those that occur commonly today in Arctic and Antarctic soils. The surprise is that they occur in huge numbers in permafrost (up to 10,000 cells per gram of dried permafrost soil). Although these are very significant numbers, they represent only 0.2% of the numbers of bacteria recovered from the same permafrost. This means that 1 gram of permafrost contains approximately 5,000,000 bacterial cells. The number of viable microorganisms as well as their diversity varied from sample to sample, but anywhere from 10 to 100,000,000 cells per gram of dried permafrost could be recovered, including bacteria, filamentous fungi, yeasts, and protists. The number and diversity of life forms present decreased with age. For example, filamentous fungi were detected only in relatively young permafrost (<10,000 ybp). However, the situation in glacial ice appears to be somewhat different, because viable fungi have been found in glacial ice dating to >500,000 ybp.

Many of the microorganisms detected in ice are almost identical to contemporary species, some resemble temperate species but with slight differences, and some are unknown to science. Abyzov detected bacteria, actinobacteria (a type of filamentous bacteria), yeasts, and filamentous fungi in the Antarctic ice sheet. The yeasts isolated from ice 700–3,250 ybp are biochemically and structurally different from temperate isolates of the same species. Differences include production of a dark pigment (melanization) as well as ultrastructural differences including an increase in mitochondrial size and number, which apparently permit Antarctic isolates to survive very long periods of anabiosis. In Abyzov's reports, filamentous fungi were not isolated from ice older than 38,600 ybp (roughly 651 m depth, 2,135 ft, in the Vostok core). We and others have cultured fungi from ice that is >500,000 years old. However, we found that some core sections yielded viable fungi, while others failed to yield viable fungi, but it was not always correlated with core section depth. Most of the fungi isolated by Abyzov are most likely native to Antarctica, because they are common microflora of the soil, air, and snow of Antarctica, and the Vostok site is high altitude and is somewhat isolated from winds that would come from regions outside of Antarctica. Isolates from other Antarctic and Greenland ice cores may be representative of a mix of species, including some from temperate latitudes. Several fungi (e.g., *Penicillium*, *Aspergillus*, *Cladosporium*, and *Mucor*) are repeatedly isolated from ice of various ages presumably because the spores of these fungi are easily windborne. In culture, the ice isolates formed spores less often than did their temperate counterparts. It is unknown whether this is an adaptive character for the polar species, or if these organisms had sustained some genetic damage during their entombment in the ice.

More than 1,000 fungal isolates (filamentous fungi and yeasts) have been cultured by several research labs (including ours) from ice cores and permafrost samples ranging in age from less than 500 to over 500,000 years old. Most of the fungi have been identified to genus and many to species. As with the bacteria, some are very similar to contemporary species, some are remotely similar, and some are so different that no one knows exactly what they are. Some are cold-loving and others appear to be temperate species. However, with most, it is difficult to determine their origins and optimal growth conditions because they have not been identified or fully characterized. Some of these taxa (e.g., *Rhodotorula*,

Cladosporium, and *Penicillium*) were recovered from many ice core sections and depths, while others were found on a single occasion within a single core section. Multiple isolations of the same species from progressively older age is feasible. This is important when examining epidemics, geological events, anthropomorphic affects, evolution, and other phenomena. Some of the taxa may be pathogenic to humans, or to animals, plants, or other organisms. The potential for danger is real, but impossible to gauge at this time.

Some ice core sections contain an abundant and diverse assemblage of microorganisms, whereas other sections (sometimes adjacent ones) contain virtually nothing. Abyzov and coworkers have used direct microscopic examination of filtered ice meltwater, and the consumption of radioactively labeled organic compounds as proof of viability rather than direct culturing on nutrient media. He was able to show that ice cores up to 240,000 years old obtained from Vostok, Antarctica, contained a wide variety of contemporary bacteria, yeasts, filamentous fungi, microalgae, pollen of higher plants, and dust particles. The filters then were stained with a fluorescent dye, which complexes with proteins, and examined by epifluorescence microscopy to quantify the number of cells present in the meltwater. SEM (scanning electron microscopy) was conducted on the filters as well to observe the size and shape of the cells. Vegetative bacterial cells, but not endospores, were detected. In addition, the number of bacteria detected was directly related to the amount of dust present in the ice, which varied by depth. Microbial cell counts ranged from 800–10,000 cells/ml of meltwater. Microbial viability decreased with increasing depth, but viable organisms were found at a depth corresponding to 240,000 ybp. When we assayed ice core sections from the Vostok ice core, the cell counts were usually between 10 and 1,000 cells/ml. In the Lake Vostok accretion ice, cell counts for some sections were as low as 1–2 cells/ml. These extremely low levels indicate an environment that supports living organisms, but at low levels.

BACTERIA IN PERMAFROST

Viable bacteria have been isolated from 3-million-year-old permafrost in Siberia, as well as 8-million-year-old ice from Antarctica. Of the many isolates recovered, dozens have been characterized and identified by sequencing a portion of their DNA and comparing these sequences to those of known bacteria. In addition to sequencing, more conventional techniques for bacterial identification including cell morphology, Gram staining, growth on specialized media, ability to produce endospores, and others were used. A great diversity of bacteria was recovered including rods, cocci, endospore formers, nonendospore formers, Gram positive, and Gram negative forms, as well as bacteria able to utilize cellulose, sulfur, and nitrogen as energy sources. While many grew only at temperatures below 22°C (72°F), and some only below 15°C (59°F), many grew above 30°C (86°F). This is a surprise, since they were found at temperatures well below freezing. More than half of the isolates were enterobacteria (bacteria that live in the gut). This was also a surprise because other reports indicated that enteric bacteria die very rapidly in Arctic soil. The origin of these permafrost enterobacteria is unknown, yet some are at least 50,000 years old. Many bacteria commonly found in Siberian soil were not found in the permafrost. Some were photosynthetic, which was surprising given the fact that light does not penetrate permafrost for more than a few millimeters. It seems apparent that only those bacteria capable of survival in a metabolically active state at -10°C (14°F), the mean temperature of permafrost) will persist because their DNA repair mechanisms remain active (albeit at very low levels) even at cold temperatures. Thus, the extreme cold, low nutrients, and lack of oxygen put strong selection pressures on the microorganisms likely to survive in permafrost. It appears that many different organisms can enter ice, but only some are capable of prolonged anabiosis, and subsequent revival after thawing.

How can bacteria found within permafrost remain metabolically active even at -10°C (14°F), the ambient temperature of permafrost? Believe it or not, there is still liquid water present within the permafrost, even at this temperature, and metabolic activity has been detected in some bacteria down to -80°C. Films of liquid water can form around soil particles and microbes even at very low temperatures. When water freezes it tends to push out solutes along the front of freezing. This increases the concentrations of solutes along the freezing front. Additionally, microbes contain high concentrations of solutes. In fact, cells are nearly crystalline. Because of these high solute concentrations, the freezing point of the accompanying water is depressed. Therefore, water freezes well below zero Celsius. Because of this, the microbes do not completely freeze and therefore they can remain metabolically active. This is important if they are to reproduce, absorb, and utilize nutrients, and repair their chromosomes and other cellular components.

Many organisms that have been found and revived from ice and permafrost are non-spore formers. Therefore, spores are not the only means by which microbes can survive stressful conditions. There have been many studies of anabiosis of bacteria that do not form spores. When these bacteria are under stressful conditions, often the cells change shape. They become smaller, presumably to condense and concentrate the cellular contents. Some bend, and many squeeze into one part of the cell, leaving the other part of the cell collapsed, and looking somewhat like an appendage or finger. Still others change very little under stressful conditions. While the ability to withstand stressful conditions (including dehydration and freezing) has generated a great deal of interest, relatively little is known about the genetic and developmental pathways leading to anabiosis.

VIRUSES

To our knowledge, we are the only investigators to have examined ice cores for viruses (although at least one other lab was looking for virus-like particles by electron microscopy), primarily because they are difficult to find in the ice. Because viruses are very small, electron microscopes must be used to search through the meltwater. This is difficult and time-consuming. To detect viable viruses, they must be grown in susceptible host cells in culture or *in vivo*. However, the type of virus must be known prior to starting these tests. RNA and DNA sequences can be more easily determined, but primers must be developed for each group or strain of virus, and this is sometimes very difficult, due to sequence variability in RNA viruses. For example, influenza A viruses have 18 types of haemagglutinin protein (designated H1 — H18) and 11 types of neuraminidase protein (designated N1 — N11), such that there are almost 200 possible combinations of these proteins in influenza. The most often circulating in humans is H1N1.

We performed a detailed examination of 30 different Greenland glacial meltwaters from three different drill sites representing 11 different ice core sections for tomato mosaic tobamovirus (ToMV) by RT-PCR (reverse transcription PCR) amplification. This virus was chosen because it is very hardy, as are all of the members of the Tobamovirus group. These plant viruses can survive space travel, and can survive in soil, water, clouds, and other environments. Infectious tobacco mosaic tobamovirus (TMV), the type member of the Tobamovirus group, has been found in cigarette ash, surviving the extremely high temperatures of burning tobacco leaves. We assumed that if tobamoviruses can survive all of these extreme conditions, certainly it could withstand being frozen in glaciers. The virus was detected in 17 of the ice core sections. We could detect as few as 150 virus particles. The virus was present in ice core sections that were from <500 to 110,000 years old. A small portion of the amplified viral genome was sequenced, and the sequences compared among ice and extant isolates of the virus. Phylogenetic analysis was performed on these sequences. Four core sections contained more than one ToMV genotype, and five core sections contained sequences nearly identical to those of extant ToMV genotypes. The results indicated that the viruses may be recycling through time. That is, old genotypes and more recent genotypes are mixing in the glacial ice, and as the mixtures melt from the glaciers, new mixtures are formed consisting of ancient genotypes, not-so-ancient genotypes, and modern genotypes. This process we call "genome recycling" and it is discussed in a later chapter. Immuno-capture PCR (IC-PCR), where virus particles are first concentrated using antibodies to the viral coat protein, corroborated the presence of ToMV in some ice melts in which virus was detected by RT-PCR, providing evidence that intact virus particles are present in the ice, an indication that they may be infectious.

We also have coaxed some prophage (viruses integrated into the host genome) from their bacterial host chromosomes, using a chemical that signals a stress response in the bacteria. *Bacillus subtilis* often carry prophage in their genomes. We used cultures of these bacteria that had been isolated from Greenland ice core sections, and stressed them to attempt to stimulate the viruses to pop out of the genome and form new virus particles. We recovered bacteriophage (bacterial viruses) from four bacterial cultures isolated from ice, again dating from <500 to approximately 110,000 years old, which formed PBSY-like phage, but the phage often appeared incompletely formed, when viewed by electron microscopy. Also, when the resultant viruses were exposed to *B. subtilis* cultures, they failed to infect the cells. We have also detected influenza A viruses in Siberian lake ice, from lakes that serve as migratory bird nurseries. Others have demonstrated that influenza A viruses can infect host cells after freezing and thawing in environmental ice and mud. It is clear that environmental ice preserves a variety of viruses.

Viruses are vital in the oceans because they are one of the major recyclers of organic matter. They do this by lysing cells, which releases large amounts of organic compounds. In areas of the ocean that have limited nutrients, this action is absolutely necessary for the organisms in those zones. This has been documented in the oceans of the world. This may be occurring in environmental ice as well. Most ice throughout the world contains low amounts of nutrients, including organic compounds. As ice freezes, it forces out a front of nutrients and organisms. As viruses lyse the cells, additional nutrients are released, creating zones of higher nutrient and microbe concentrations. Although it is generally thought that viruses are damaging parasites, they are actually crucial components of many ecosystems, and seem especially important in zones of low nutrient concentrations. Viruses may also be vital in ice and subglacial ecosystems.

SOURCES AND ADDITIONAL READINGS

Abyzov, S.S., M.N. Poglazova, J.N. Mitskevich, and M.V. Ivanov. 2005. In: Castello, J.D. and S.O. Rogers (eds.) *Life in Ancient Ice.* Princeton NJ: Princeton University Press, pp. 240-250.

Bidle, K.D., S. Lee, D.R. Marchant, and P.G. Falkowski. 2007. Fossil gene microbes in the oldest ice on Earth. *Proc. Natl. Acad. Sci. USA* 104:13455-13460.

Castello, J.D., and S.O. Rogers. 2005. *Life in Ancient Ice.* Princeton, NJ: Princeton University Press.

Castello, J.D., S.O. Rogers, J.E. Smith, W.T. Starmer, and Y. Zhao. 2005. Plant and bacterial viruses in the Greenland ice sheet. In: Castello, J.D., and S.O. Rogers (eds.) *Life in Ancient Ice.* Princeton NJ: Princeton University Press, pp. 196-207.

Castello, J.D., S.O. Rogers, W.T. Starmer, C. Catranis, L. Ma, G. Bachand, Y. Zhao and J.E. Smith. 1999. Detection of tomato mosaic tobamovirus RNA in ancient glacial ice. *Polar Biology* 22: 207- 212.

D'Elia, T., R. Veerapaneni, and S.O. Rogers. 2008. Isolation of microbes from Lake Vostok accretion ice. *Appl. Environ. Microbiol.* 74: 4962-4965.

D'Elia, T, R. Veerapaneni, V. Theraisnathan, and S.O. Rogers. 2009. Isolation of fungi from Lake Vostok accretion ice. *Mycologia* 101: 751-763.

Ma, L., C. Catranis, W.T. Starmer, and S.O. Rogers. 1999. Revival and characterization of fungi from ancient polar ice. *Mycologist* 13: 70-73.

Ma, L., S.O. Rogers, C. Catranis, and W.T. Starmer. 2000. Detection and characterization of ancient fungi entrapped in glacial ice. *Mycologia* 92: 286-295.

Rivkina, E., J. Laurinavichyus, and D.A. Gilichinsky. 2005. Microbial life below the freezing point within permafrost. In: Castello, J.D., and S.O. Rogers (eds.) *Life in Ancient Ice.* Princeton NJ: Princeton University Press, pp. 106-117.

Rogers, S.O., Y.M. Shtarkman, Z.A. Koçer, R. Edgar, R. Veerapaneni, and T. D'Elia. 2013. Ecology of subglacial Lake Vostok (Antarctica), based on metagenomic/meta transcriptomic analyses of accretion ice. *Biology* 2: 629-650.

Shtarkman Y.M., Z.A. Koçer, R. Edgar R, R.S. Veerapaneni, T. D'Elia, P.F. Morris, and S.O. Rogers. 2013. Subglacial Lake Vostok (Antarctica) accretion ice contains a diverse set of sequences from aquatic, marine and sediment-inhabiting Bacteria and Eukarya. PLoS ONE 8(7): e67221. doi:10.1371/journal.pone.0067221.

Vishnivetskaya, T.A., L.G. Erokhina, E.V. Spirina, A.V. Shatilovich, E.A. Vorobyova, A.I. Tsapin, and D.A. Gilichinsky. 2005. Viable phototrophs: cyanobacteria and green algae from the permafrost darkness. In: Castello, J.D., and S.O. Rogers (eds.) *Life in Ancient Ice*. Princeton NJ: Princeton University Press, pp. 140-158.

LINKS

https://cosmosmagazine.com/palaeontology/big-five-extinctions
https://curiosity.com/topics/a-russian-scientist-injected-himself-with-35-million-year-old-bacteria-curiosity/
https://en.wikipedia.org/wiki/Extreme_weather_events_of_535%E2%80%93536
https://www.newscientist.com/article/dn7064-ice-age-bacteria-brought-back-to-life/

15 Pathogens in the Environment: Viruses

"Watch out where the huskies go, and don't you eat that yellow snow"

Frank Zappa

EXPOSURE TO PATHOGENS

Based on the numbers of viable microbes that have been reported in environmental ice (e.g., glaciers, ice domes, ice shelves), and the rates of ice melting throughout the world, we estimate that during an average year, approximately 10^{18} to 10^{22} viable microbes (including fungi, bacteria, and viruses) are released from melting ice. Sea ice may contribute similar, or larger, amounts. Oceans are in direct contact with sea ice, as well as with the margins of the world's glaciers. A large proportion of the human population, as well as animals, drink water that has melted from glaciers. Ocean currents and winds can spread microbes worldwide, ensuring that each of us (and most other organisms on Earth) is in direct, daily contact with them. Additionally, animals can carry viruses and other microbes throughout their movements and migrations. For example, Arctic Terns migrate from the Arctic to the Antarctic, and back again, an average distance of 71,000 km (44,000 mi). Filter feeders (e.g., clams, scallops, oysters) and parasites also may concentrate microbes that increases the likelihood of receiving infectious doses of any pathogens present. Wind and surface storms aerosolize water droplets that may carry large numbers of microbes.

Our exposure to pathogens is on the rise because of increased melting of ice worldwide, as well as increases in world travel and population. We all are in intimate contact with microbes thawing from environmental ice sources. They are on us and they are in us. Glaciers worldwide are receding at unprecedented speeds. For example, in the Alps, some trail markers that once were at the height of hiker's eyes now are tens of meters above their heads, because the glaciers below the markers have been reduced, or have disappeared. In Antarctica, large chunks of the ice shelves are shrinking, because large pieces have broken off. Much of the Larsen B ice shelf has collapsed into the ocean. Almost 30,000 km² (20,000 mi²) separated from the ice shelf. This type of erosion of the polar ice is increasing, delivering microbes to all parts of the globe, with a potential for disease and mortality much greater than any war or other natural disaster (with the exception of a large asteroid collision with Earth). It remains to be seen whether the rapid melting will have devastating consequences for the Earth and its inhabitants.

INFLUENZA A

Although the 1918 influenza pandemic is often colloquially called the "Spanish Flu," its origins were far from Spain. Much of the epidemiological evidence instead points to Asia as the origin. In 1917, there was an outbreak of a respiratory illness in Northern China that turned out to be identical to what would eventually be called the Spanish Flu, or the 1918 flu. By the end of 1917, it had spread more than 300 miles across China, killing dozens of people daily. Also, in 1917, 25,000 Chinese laborers arrived in Canada, of which more than 3,000 had flu-like symptoms that also turned out to be influenza. In January of 1918, France and Britain imported 96,000 Chinese laborers to work behind the front lines of the Western Front during World War I. They wanted their trained soldiers on the front lines, so they hired Chinese workers to do those other tasks behind the lines. The Chinese laborers were shipped to England and France in crowded rail cars, under poor sanitary conditions. Many were sick upon arrival and were sent to hospitals in France, where large numbers of them died. By March of 1918, the flu had become more deadly, and had spread into Africa and from coast to coast in the US. At Camp Funsten, Kansas, 48 soldiers died from the disease that year. It continued to infect and kill people across the globe. Eventually, it killed more than 50 million people worldwide, until it finally waned and disappeared by 1920. Where did it come from? More importantly, where did it go, and will it return?

China has been the origin of many strains of influenza A for several reasons, and the origin of the deadly 1918 flu followed a standard pattern. China is on the flyways for many types of migratory waterfowl, all of which potentially can carry strains of influenza A virus. Several years ago, we demonstrated that influenza A viruses are present in frozen Siberian lakes. Other researchers have found viable influenza A viruses in ice and mud from ponds and lakes. When the birds migrate north in the Spring to mate and hatch their chicks, the ice in the lakes is thawing and releasing infectious viruses from bird feces deposited during the previous Summer and Fall. The waterfowl drink the water, ingesting the viruses. The infected birds release virus-containing waste into the lakes, which is frozen in the surface ice during the Fall and Winter, to be released again the next Spring, in the process known as genome recycling (discussed in a later chapter). But, the birds arriving and becoming infected early in the Spring are not the same birds that were the last ones to leave in the Fall. So, the mix of viruses differs from season to season. Influenza A virus strains in the Northern Hemisphere usually are first observed in the north, then they proceed southward. This is because the birds fly south during the Fall and Winter months, and many may carry influenza A viruses. These can be transmitted to domestic birds, pigs, and people, which are regularly in close proximity to one another in Chinese houses, farms, and markets. There, the virus strains can combine within the animals to produce new strains through genetic recombination and chromosome reassortment. This is the process whereby each of the eight chromosomes of influenza A can be randomly distributed into newly produced virus particles, providing a huge number of possible combinations of chromosomes and genes.

Influenza A viruses have eight chromosomes, and when infected by more than one strain, the variants of the chromosomes can mix to produce unique virus particles. Many strains carry virus particles that have a mixture of chromosomes from human strains, some from pig strains, and some from bird strains. In fact, the 1918 influenza A virus contained some pig, some bird, and some human virus chromosomes. Some chromosomes also showed signs of recombination among the types of chromosomes. That is, one of the chromosomes appeared to be mainly from pig hosts, with a few regions being closer to a bird sequence, indicating a prior recombination event. Another was mainly from a human strain, but with some regions appearing closer to a bird strain, and yet another had a chromosome from a bird source, with a few regions looking more like a human strain. Each of the recombinant chromosomes had been in different combinations within different strains, but when they were brought together in 1917, the strain became much more deadly, partly due to an extreme immune response in the people.

The 2009 influenza A strain that caused high rates of infection, spread, and death also was a reassortant strain. It had only one chromosome from a human strain, two from birds, and five from pigs. This type of genetic mixing can make the virus unrecognizable to the immune system, and thus the virus can infect and reproduce in the host organisms (e.g., people), and sometimes this can induce extreme immune responses that lead to rapid decline and death in patients, as occurred in the 1918 pandemic. This influenza A strain was unusually deadly because it caused the lungs of the patients to fill rapidly with fluids, which was part of the immune response, and the people essentially drowned, sometimes only two days after first feeling ill. Influenza A survives freezing and thawing well, and also has been found in lake and pond ice, cold water, and mud. There is a good possibility that some influenza A strains are frozen in polar and glacial ice. If the 1918 influenza A returns, the world population will be susceptible to it, which could trigger another deadly pandemic.

Another possibility for a pandemic recently has been put forward. Bats carry influenza A, but the strains differ from those of influenza A carried by birds, pigs, chickens, and other organisms, in that the virus particles attach to MHC II (major histocompatibility class II) host cell receptors instead of sialic acid receptors that are the usual sites of virus attachment. Both are cell surface receptors, used by cells to identify one another. While MHC II receptors are variable among species, they are more conserved than sialic acid receptors, which implies that influenza A carried by bats can more readily move from one species to another. This can cause serious and deadly disease. Because it has not been documented to have occurred in the recent past, the immune systems of humans worldwide have never been exposed to the proteins in this virus, and thus are completely naive to this virus. In other words, everyone on Earth will be susceptible to these strains of influenza A viruses. Species of bats have distributions that range from the Arctic, across the equator, and into the Antarctic. This includes areas of dense human populations. Bats make up 20% of the total mammal population. That is, if you count all of the mammals in the wild, one out of five of those individuals is a bat. This presents a very troubling scenario, because a strain of influenza A could be spread rapidly around the world in bat populations, and if it causes immune reactions similar to the 1918 influenza A strain, many more than 50 million people could succumb to the infection.

OTHER VIRUSES

The majority of viruses survive freezing and thawing. Thus, the potential for their release from ice, and infection of susceptible hosts appears to be high. The most likely candidates would be those in greatest abundance, distributed by wind (and/or water), and survive well when frozen and then thawed repeatedly. There are many candidates, and we have already discussed influenza A virus, primarily because of its history of killing millions of people. Caliciviruses are very hardy, and many also are dangerous pathogens, and therefore, pose another potential candidate for emergence from melting ice. These viruses, including norovirus, Norwalk-like virus, and related viruses, infect sea animals and land mammals, including swine, humans, horses, and others. The Norwalk-like virus that causes many outbreaks of severe human diarrhea (and other disease symptoms) annually is highly infectious, and has mortality rates of 15–20%. A similar virus in cats has a mortality rate of nearly 70%. They also can cause spontaneous abortions in sea mammals, some land mammals, and possibly in humans. Evidence of caliciviruses has been found in the blood of patients with hepatitis at higher rates than in the general public. From 15% to 40% of women who had spontaneously aborted fetuses tested positive for caliciviruses. Only 2 to 5% of the general public tests positive for the same viruses. Some have very broad host species ranges; including some that can infect fish, seals, swine, rabbits, cattle, primates, felines, canines, and humans. Disease symptoms range from diarrhea to hepatitis to pneumonia.

California Gray Whales migrate annually from the Sea of Cortez to the Eastern Siberian Sea, swimming past the coasts of Alaska, Canada, the US mainland, and Baja, Mexico, and into the Sea of Cortez. Meltwater from glaciers and sea ice mix with the seawater in many parts of the ocean along the way. These whales are continuously exposed to a variety of caliciviruses. They pass populations of sealions, fur seals, harbor seals, and gray seals, all of which also are repeatedly exposed to many types of calicivirus. Frequent movements of caliciviruses into terrestrial animals from aquatic species suggest active transport mechanisms for caliciviruses from sea animals to land mammals. It has been suggested that aerosolized sea water containing viruses falls on plants that are then ingested by terrestrial animals. Some caliciviruses also are able to infect fish, thus extending the distribution of these prolific viruses. Caliciviruses are very hardy. They survive for weeks in seawater, even up to 15°C (59°F). They probably survive for very long periods of time in ice (although the exact durations have not yet been determined). They occur in high concentrations in sea mammals, seawater, and sea foam. Thus, virus particles could easily be blown from the oceans into the air and eventually deposited in glacial ice or in locations where terrestrial animals might come into contact with them.

Several incidences of calicivirus appearance, disappearance, and reappearance, with decades between each, have been documented. In the 1930's, there was an outbreak of a calicivirus strain in swine. Over the next twenty years it appeared frequently, even with extensive eradication measures. Then, suddenly, it completely disappeared. A dozen years later, it reappeared in other types of animals, including fish. Had it come back through the melting of reserves in ice? No one knows. Another example is the emergence of a related calicivirus in swine in New Jersey. Just as suddenly, it disappeared. Then, twenty-four years later, it appeared in whales in Alaska, and sealions in California. Again, had the animals been exposed to ice meltwater, or was the mechanism something else? No one knows, but ice is a possibility, and no other plausible mechanism has been suggested. The general pattern for nearly all calicivirus types is that they tend to disappear from a given population from one year to the next. If they do reappear, it is usually years later. This phenomenon could be explained by periodic reinfection of target species with viable viruses released from ice.

ICE AS A VIRUS RESERVOIR

When we detected viruses in Greenland ice that was more than 100,000 years old, the question most often asked of us by the news media was, "Can pathogenic viruses melting from glaciers cause human epidemics?" The short answer is yes, but the probability is low. The second question was, "Have they caused outbreaks and epidemics in the past?" And, the answer is, probably yes, but it is unknown at this time. The majority of viable microbes, including viruses, that have been isolated from glaciers are not pathogens. Most that survive being frozen in a glacier are adapted to temperatures below the normal range for most animals. However, many can grow at higher temperatures. They are adaptable organisms that can survive freeze/thaw cycles, and are adept at growing under a number of conditions. In short, they are very hardy organisms. Some can live on plants and animals, but even most of those are not pathogenic. Therefore, the probability of an epidemic or pandemic arising from a pathogen melting from a glacier is low, but certainly not zero. For example, many pathogens can be frozen and stored in a freezer, and when thawed retain their infectivity. Influenza A viruses retain their infectivity after being frozen and thawed many times. This has been demonstrated in laboratories as well as under environmental conditions. Likewise, many viruses, bacteria, fungi, amoebae, and protists survive freezing and thawing.

Disease outbreaks in animals, humans, and plants caused by unknown or newly discovered microbes are commonly reported. Where do they come from? Why haven't they been observed previously? Are they from biotic reservoirs? Are they mutant strains of unknown pathogens? Or are they from abiotic reservoirs? Other candidate viruses include some of the enteroviruses (e.g., polioviruses, echoviruses, rotaviruses, and Coxsackie viruses). Polioviruses (as well as all enteroviruses) spread in feces-contaminated water, and may become airborne in water droplets. These viruses remain viable for long periods of time when frozen. Currently, there is a worldwide effort to eradicate poliovirus. However, if it survives in glacial ice, this will be impossible. One promising aspect for eradication of this virus is that it is an RNA virus, meaning that its genome is more susceptible to damage, and therefore long-term survival is lower than that of a DNA virus. Eradication would be deemed successful if it does not reappear decades or centuries after poliovirus is eradicated from all individuals in the human population. Time will tell.

Rotavirus, another RNA virus, also survives repeated freezing and thawing events well. It causes severe diarrhea, and is widespread, resulting in hundreds of thousands of deaths annually, especially in children. Infection rate peaks move from south to north in the Northern Hemisphere (the reverse occurs in the Southern Hemisphere). This coincides with cool dry weather. The virus is mobilized in cool water as snow and ice melting occurs, as well as on dust particles in the air by inhalation. It can be contracted directly from ice (including ice cubes and ice chips in drinks) or by drinking contaminated cold water. Recently, researchers have connected rotavirus infections with the development of type 1 diabetes, which has been on the increase in recent years. There is a genetic component in the mix, as well, because the highest incidence of type 1 diabetes is in those from northwest Europe, and their descendants. This primarily includes people from Europe, North America, and Australia. Whether infections are contracted from melting ice is yet to be determined, but the pattern of appearance and spread makes it a possibility.

Many other viruses can retain infectivity after one or more freeze/thaw cycles, including poxviruses, measles virus, herpes virus, yellow fever virus, tobacco mosaic virus, tobacco necrosis virus, cauliflower mosaic virus, and many other animal and plant viruses. Recently, an RNA virus and a DNA virus were found and sequenced from 700-year-old caribou feces extracted from permafrost. The RNA virus was a cripavirus that infects insects, while the DNA virus was a geminivirus that infects plants. The researchers could not demonstrate infectivity by the cripavirus, partly because the host was uncertain, but the geminivirus was used to successfully infect tobacco plants. This means that viruses frozen in the permafrost around 1300 CE are still viable today. Smallpox is a DNA virus that was eradicated a few decades ago, although some stocks of the virus exist in a few secure labs. It has existed in human populations for at least 12,000 years, causing many deadly epidemics. Some Egyptian mummies have signs of the disease. Although it has hopefully been eradicated, smallpox virus can be frozen and thawed, while retaining its infectivity. So, it may possibly reemerge from a frozen source somewhere in the world. The human population would be very susceptible to this disease because vaccination programs ended long ago. There also are probably many viruses unknown to science that may have the capability to infect humans. Viruses such as Ebola, HIV, and other similar viruses are unlikely to be preserved in environmental ice because they usually are transmitted through body fluids, do not survive for long outside of their hosts, and are difficult to preserve by freezing.

ICE CUBES TO GLACIERS

As mentioned in an earlier chapter, our trek into the study of organisms in ancient ice began at the urging of a Syracuse University colleague. He had been working with ice core sections from Greenland for about 5 years, but got the idea for this research some years earlier. He and his friend from the University of Miami, were in, of all places, the Caribbean. They ordered drinks, which came with the obligatory ice cubes. They wondered aloud whether there were any disease-causing organisms in the ice and how long they might survive in the ice. This question has been answered for many pathogenic microbes, and the answer is, yes, many of them, including cholera and rotavirus, can be transmitted through ice cubes and ice chips. If they survive there, could they survive in glacial ice? By the way, fortunately, neither of them got sick, probably because the ice contained few pathogens, and they had good immune systems. When they related this story to us, I (SOR) remembered that my 8th grade Earth Sciences teacher had said that you could get terrible diarrhea from drinking glacial meltwater. He said that the diarrhea was caused by high concentrations of certain minerals, primarily magnesium, in the glacial ice water. But, maybe the minerals are not the only cause. Many microbes also cause diarrhea, as well as other maladies. This was our embarkation into the study of microbes in ancient ice. Our first questions dealt with the characteristics of ice. How and why can microorganisms survive in ice? Two other questions whose answers still elude us are: How do organisms remain alive in the ice for extended periods of time (thousand to millions of years)? and, how do they survive the very high pressures caused by the weight of the ice, which can be kilometers thick?

DON'T EAT THAT ICE

The diversity of microbial life on Earth truly is astounding. There are perhaps over 5 million species of microbes on Earth, but only about 100,000 have been described. Equally surprising is the number of organisms that survive and thrive in and around ice. People have used ice to retard food spoilage for thousands of years. Until the first quarter of the 20th century, ice was used routinely to treat wounds, primarily to reduce swelling. Thus, the tacit assumption was that nothing survives or grows in ice. This was a false assumption, and many people became infected through contact with ice meltwater. Some had limbs amputated, and others died due to the ice treatments! While some organisms are killed by freezing and thawing, we now know that others easily survive these conditions. Some even grow in ice, and others are routinely frozen as a part of their normal life cycle. Organisms as large as scorpions, frogs, and turtles can survive being frozen and thawed. Perhaps one day humans can be successfully frozen and thawed, but for now realize that multitudes of organisms are melting out of glaciers daily, and some are even melting out of the ice cube in your drink. As was stated earlier, roughly 10^{18} to 10^{22} microbes are released from melting ice annually. However, only a small percentage are pathogenic, and an even smaller number are human pathogens. Some of them do not survive freeze-thaw cycles very well. But, even assuming that this brings the number of pathogens that could be released from melting ice to 10^{10} to 10^{16}, these organisms still pose potential risks for human populations. It is still likely that an outbreak, epidemic, or pandemic will occur from some pathogen melting from environmental ice, and it is likely that this has occurred many times in the past.

DETECTING VIRUSES EMERGING FROM ICE

The list of microbes that can survive freezing and thawing is large and diverse, and the number of pathogenic microbes is a subset of this. Those that can infect humans is a subset of that subset. However, there are some significant pathogens capable of creating an outbreak, epidemic, or pandemic. As mentioned previously, anthrax can be preserved in permafrost and ice, and remains infectious after it thaws. Influenza A virus strains also survive environmental freeze-thaw cycles. Constant surveillance is required to avoid epidemics and pandemics. Even with surveillance and vaccination, influenza A kills between 20,000 and 50,000 people in the US annually. Worldwide, more than half a million deaths have been attributed to influenza A yearly. Currently, it is unknown how much of the annual circulating virus is the result of thawing of ice and ice-covered lakes, but it could be substantial.

Viruses differ from cellular life forms in that they have a simple, acellular organization, and they require a cellular host for their reproduction and spread. Viruses are classified on the basis of their nucleic acid type and characteristics. Viral genomes may be one or several different molecules of RNA or DNA (including single or double stranded, plus or minus or ambisense, linear, closed circle, or either one), virion size and shape, the type of host they infect (plant, animal, bacteria, archaea, fungi, protists), the symmetry of their protein coat, intracellular location of viral replication, mechanism of transmission, and type of disease induced, as well as other features. Viruses are responsible for influenza, smallpox, chickenpox, measles, Ebola, herpes, cancer, and many other diseases. While these viruses can cause devastating diseases, most other viruses are not harmful, and we hardly notice them. In many ecosystems, they are absolutely vital to the recycling of nutrients. Also, they have been extremely important in the evolution of life on Earth, in that they serve as agents to move genetic elements from one organism to another with great frequency and speed. Humans and other organisms would not have evolved without viruses. Many bacterial and fungal species cannot survive without them.

We have been engaged in developing assays for other viruses (e.g., primers and probes for polioviruses, echoviruses, coxsackieviruses, influenza viruses, poxviruses, rotaviruses, canine and porcine parvoviruses, herpesviruses, caliciviruses, potato virus X, tomato bushy stunt viruses, and additional tobamoviruses). These viruses were selected because they share one or more of the following characteristics: waterborne, airborne, geographically widespread, and/or highly stable, (i.e., resistant to degradation outside of a living host cell). Poxviruses were selected because in 1979 the World Health Organization (WHO) declared that smallpox virus was eradicated from the face of the Earth (the only life form ever to be deliberately exterminated from the Earth). We thought it would be interesting to determine if some of these viruses might still exist in ancient ice. To date, we have not found any poxviruses in any ice core sample, but we continue to look. This brings up one of the hazards of working with these ice cores. If we do find human pathogens, there is a chance that they are still infectious. While most older people have received smallpox vaccinations, most younger people have not, since the belief is that it no longer exists on Earth (except for strains retained in a few laboratories). We sincerely hope that this is true. We have also used metagenomic and metatranscriptomic methods to determine all of the nucleic acids in an ice core section. While few viruses have been detected using these methods to date, almost half of the data so far collected has not been found to correspond to anything on the national and international sequence databases. It is possible that most of these are virus sequences, but because virus genomes are so variable, and few scientists study viruses, their representation on these databases is small.

While we should be concerned about known pathogens, there are probably many more that we don't know anything about. As a senior US administrator once said, "What we don't know, we don't know." Our surveillance systems for known diseases are imperfect and miss many disease outbreaks, and surveillance efforts focus generally on host reservoirs rather than abiotic reservoirs. Our surveillance for unknown diseases is nonexistent. Needless to say, many threats to humans (and other organisms) are completely ignored. We suggest that surveillance of ice is advisable. Presently, the human population is rapidly growing, and the melting of ice worldwide is increasing. These two phenomena may have more importance to the future of humans on Earth than anything else.

SOURCES AND ADDITIONAL READINGS

Castello, J.D., and S.O. Rogers. 2005. *Life in Ancient Ice*. Princeton, NJ: Princeton University Press.

Castello, J. D., S. O. Rogers, J. E. Smith, W. T. Starmer, and Y. Zhao. 2005. Plant and bacterial viruses in the Greenland ice sheet. In: Castello, J.D., and S.O. Rogers (eds.) *Life in Ancient Ice*. Princeton NJ: Princeton University Press, pp. 196-207.

Castello, J.D., S.O. Rogers, W.T. Starmer, C. Catranis, L. Ma, G. Bachand, Y. Zhao, and J.E. Smith. 1999. Detection of tomato mosaic tobamovirus RNA in ancient glacial ice. *Polar Biology* 22: 207- 212.

D'Elia, T., R. Veerapaneni, and S.O. Rogers. 2008. Isolation of microbes from Lake Vostok accretion ice. *Appl. Environ. Microbiol.* 74: 4962-4965.

D'Elia, T, R. Veerapaneni, V. Theraisnathan, and S.O. Rogers. 2009. Isolation of fungi from Lake Vostok accretion ice. *Mycologia* 101: 751-763.

Ma, L., C. Catranis, W.T. Starmer, and S.O. Rogers. 1999. Revival and characterization of fungi from ancient polar ice. *Mycologist* 13:70-73.

Ma, L., S.O. Rogers, C. Catranis, and W.T. Starmer. 2000. Detection and characterization of ancient fungi entrapped in glacial ice. *Mycologia* 92: 286-295.

Rogers, S.O., Y.M. Shtarkman, Z.A. Koçer, R. Edgar, R. Veerapaneni, and T. D'Elia. 2013. Ecology of subglacial Lake Vostok (Antarctica), based on metagenomic/meta-transcriptomic analyses of accretion ice. *Biology* 2: 629-650.

Rogers, S.O, W.T. Starmer, and J.D. Castello. 2004. Recycling of pathogenic microbes through survival in ice. *Med. Hypoth.* 63: 773-777.

Shtarkman Y.M., Z.A. Koçer, R. Edgar R, R.S. Veerapaneni, T. D'Elia, P.F. Morris, and S.O. Rogers. 2013. Subglacial Lake Vostok (Antarctica) accretion ice contains a diverse set of sequences from aquatic, marine and sediment-inhabiting Bacteria and Eukarya. PLoS ONE 8(7): e67221. doi:10.1371/journal.pone.0067221.

Smith, A.W., D.E. Skilling, J.D. Castello, and S.O. Rogers. 2004. Ice as a reservoir for pathogenic animal viruses. *Med. Hypoth.* 63: 560-566.

LINKS

https://en.wikipedia.org/wiki/Bird_migration
https://en.wikipedia.org/wiki/Influenza_A_virus
https://en.wikipedia.org/wiki/Larsen_Ice_Shelf
https://en.wikipedia.org/wiki/Spanish_flu
https://www.sciencemag.org/news/2014/10/virus-resurrected-700-year-old-caribou-dung
https://www.sciencedaily.com/releases/2019/02/190220133534.htm

16 Pathogens, Hazards, and Dangers

"Time wounds all heals"

John Lennon

PATHOGENS

Many pathogenic bacteria, eukaryotes, and viruses survive being frozen for long periods of time. When the surrounding temperature is between 0°C (32°F) and -15°C (5°F), many microorganisms form microlayers of liquid water around them, due to high concentrations of chemicals produced by the microbes. This protects them from damage caused by the formation of ice crystals. At -40°C (-40°F) and lower, ice freezes spontaneously outside the microbial surfaces, forming very small crystals. To some extent, ice crystals can form inside microbes, but if too many crystals form, or if large crystals form, the cell structures are disrupted and the cell may die. However, because of the high solute concentrations inside microbes, often "controlled" freezing occurs, which causes little damage. Damage can occur if the crystals form slowly and grow large, but many species produce antifreeze proteins that cause only small crystals to form around these proteins, thus controlling the size of the ice crystals. In general, Gram negative bacteria are less resistant to freezing than Gram positive bacteria. Spores of bacteria and fungi are mostly resistant to freezing and desiccation, although many non-spore forming bacteria and fungi survive freezing as well. Cells are in the most danger of dying during the freezing and thawing process. Once frozen, and held in a frozen state, the death rate decreases greatly. The rate can be further reduced by lowering the temperature to below -70°C (-94°F). Studies of eight major groups of human pathogenic viruses indicated that members of only two groups were significantly affected by freezing. The others either were slightly affected at some temperatures, or not affected at all. Pathogenic viruses and cellular organisms, as well as organisms related to pathogens, have been isolated from ice. Many bacteria that are related to anthrax have been found in the ice from Greenland and Antarctica. Anthrax (*Bacillus anthracis*), itself, has been isolated from permafrost. Several types of fungi that are potentially pathogenic (on plants and animals, including humans) also have been isolated from polar ice and permafrost. We isolated *Ulocladium atrum*, a broad range plant pathogen, from Greenland glacial ice. It is highly resistant to freezing, desiccation, heat, and ultraviolet irradiation. It is expected that it could survive in ice for hundreds of thousands of years, or more. *Alternaria* and *Penicillium* species, many of which are plant pathogens, have been isolated many times from ice core sections. Human pathogenic species of *Aspergillus* also have been isolated from ancient ice core sections from Greenland and Antarctica. Bacteria, eukaryotes, and viruses all have been found in glaciers, subglacial ice, surface lake ice, and sea ice. Therefore, clearly, meltwater from environmental ice, especially now that melting has drastically increased due to increases in greenhouse gases, poses a threat to plants and animals (including humans), and other groups of organisms.

BACTERIA

Although bacteria in many genera have been detected in glacial ice, permafrost, subglacial lake ice, surface ice, and sea ice, members of the Alphaproteobacteria, Betaproteobacteria, Gammaproteobacteria, Actinobacteria, and Firmicutes are among the most frequently detected. Members of these phyla are usually found in soil or in water, and grow under diverse conditions. Species and strains include those that are beneficial, as well as pathogens. One of the reasons that these bacteria are found still alive in ice and permafrost is that many are able to form endospores, which are thick-walled cells that are resistant to heat, cold, desiccation, and other harsh conditions. Viable endospores have been recovered from the intestines of bees encased in amber for more than 25 million years. Some microbiologists speculate that they may even be capable of surviving the harsh conditions of interstellar space for millions of years. Nevertheless, they are extremely important from a survival standpoint for those bacteria that produce them, because they are produced when environmental conditions become limiting such as during droughts or nutrient depletion. Although endospore-forming bacteria are widely distributed, most primarily inhabit soils. This may explain why such bacteria are so commonly isolated from ancient ice and permafrost. Permafrost is permanently frozen soil. When soil that contains endospores dries and becomes airborne, the endospores can be carried long distances by wind, deposited onto glaciers in which they become buried, and then released when the ice that contains them thaws.

Many types of pathogenic bacteria survive freezing and thawing, including *Bacillus anthracis* (causative agent of anthrax), *Clostridium botulinum* (botulism), *Clostridium tetani* (tetanus), *Rickettsia prowazekii* (typhus), *Mycobacterium tuberculosis* (causative agent of tuberculosis), strains of *E. coli* (including some that are deadly), and hundreds of others. However, some are sensitive to freezing and thawing. *Yersinia pestis* (plague) can sometimes survive freezing and thawing, but it is unlikely to be preserved in environmental ice for any length of time, and even if it were preserved, a fairly large number of cells would have to be released from melting ice to be infectious. Usually, with any of these microbes, more than one infectious particle or cell is needed to cause an infection, and often hundreds have to be present to successfully infect a host organism. Also, the propagules would need to have been preserved, and released into populated areas. The plague spread throughout Europe and Asia (as well as other areas) several times in the past, and some of the bacterial cells may have been frozen in the glaciers throughout parts of Europe and Asia. It also can still be found in animals throughout the world, such that there is a living reservoir of this pathogen. Its pattern of epidemics followed by disappearance suggests that there might also be a frozen reservoir.

Many species of *Bacillus* (in the phylum Firmicutes) are of considerable importance. Many produce antibiotics. *Bacillus cereus* can cause food poisoning, and infect humans. *Bacillus anthracis* is the causal agent of anthrax, a disease of livestock, wild mammals, and humans, and a potential biological weapon. Several species, such as *B. thuringiensis*, are used as bioinsecticides, as well as to protecting crop plants from freezing. Many Firmicute species in the genera *Leuconostoc*, *Streptococcus*, *Clostridium*, and *Streptococcus* are pathogens. They cause diseases, such as strep throat, rheumatic fever, food poisoning, boils, abscesses, toxic shock syndrome, pneumonia, botulism, tetanus, gas gangrene, and more than 100 other important diseases. While many of the species in this *Clostridium* are pathogenic (e.g., *C. tetani*, *C. perfringens*, and *C. botulinum*), many species are major parts of human intestinal microbiota.

Actinobacteria also have been found in ice. Most actinobacteria produce filaments (called hyphae) and asexual spores that are not endospores. In overall morphology they resemble fungi more than bacteria. The spores of actinobacteria are not heat resistant like endospores but they can survive desiccation and freezing very well, which probably permits them to survive in ice in which they have been found. These bacteria have considerable practical importance. They are widely distributed in soil where they degrade an enormous variety of organic compounds and thus function in the mineralization of organic matter. Actinobacteria also produce most of the antibiotics used in medicine today including chloramphenicol, erythromycin, neomycin, streptomycin, tetracycline, nystatin, amphoteracin B, and related therapeutic drugs. Several actinobacteria are human, animal, and plant pathogens. The bacteria that cause leprosy (*Mycobacterium leprae*), tuberculosis (*Mycobacterium tuberculosis*), and diptheria (*Corynebacterium diphtheriae*) in humans, lumpy jaw (*Actinomyces bovis*) in cattle, and potato scab (*Streptomyces scabies*) disease are actinobacteria. Several actinobacterial genera have been found frequently in ice including species of the genera *Rhodococcus*, *Nocardia*, *Micrococcus*, and *Streptomyces*. The genus *Thermoactinomyces* (in the phylum Actinobacteria) contains bacteria that are filamentous and are thermophilic (i.e., grow at high temperatures 45–60°C, and produce heat-resistant endospores that can survive a 30-minute exposure to 90°C). Although it might seem counterintuitive to find thermophiles and thermotolerant species in environmental ice, we and others occasionally find these in the ice samples. They are likely to be spread by wind currents from volcanos, hot pools, or other hot areas. Spores of *T. vulgaris* were recovered from the mud of a lake and were still viable after 7,500 years of dormancy. There are many other genera of bacteria within this group that are indispensable to the food and dairy industry because they are used to make pickles, sauerkraut, beer, wine, juice, sourdough bread, cheese, and yogurt. Additionally, species of *Nocardia* (Actinobacteria) are pathogens of various animals, including humans, that infect several organs, including lung and brain tissues causing serious diseases.

Proteobacteria is the largest and most diverse bacterial phylum. There are five major classes (Alpha-, Beta-, Gamma-, Delta-, and Epsilonproteobacteria), as well as three minor classes (Acidithiobacillia, Hydrogenophilailia, and Oligoflexia). Members of all five major classes have been found in ice and subglacial lakes, although Alphaproteobacteria, Betaproteobacteria, and Gammaproteobacteria are most frequently found in ice and subglacial lakes. Other than *Escherichia coli*, the familiar bacterium found in human (and other animal) intestines, *Pseudomonas* is one of the more recognized members of this group. Some of the pseudomonads are pathogenic, some produce a diffusible, fluorescent green water-soluble pigment, and some produce unique storage products. Several species of this genus are true psychrophiles (cold-loving), which can spoil refrigerated milk, meat, fish, and eggs. This explains one reason why they are commonly detected in ice. Some species are important pathogens of animals and plants. *Pseudomonas aeruginosa* causes infections in people with compromised immune systems, while *P. syringae* is an important plant pathogen.

ARCHAEA

Archaea comprises one of the three domains (or empires) of all cellular living things on Earth. Although they are superficially similar to bacteria, in that they are unicellular and microscopic, they differ in many ways from bacteria, and in an evolutionary sense, they are closer to eukaryotes than to bacteria. The cell wall structure and chemistry, membrane lipids, molecular biology, genomes, and metabolism are unique to this group. Because of their special metabolic adaptations, and the unique chemistry and structure of their walls and cell membranes, some archaea are capable of growing in extreme environments. Many species grow in specialized habitats, e.g., anaerobic, hypersaline, low temperatures, high temperatures. They were first discovered growing around hot thermal and hypersaline pools, such as those found in Yellowstone National Park and in New Zealand. They have also been found living near deep sea thermal vents. Species of Archaea include those that use methane and those that produce methane, so they are important in methane cycling. They have also been discovered in cold environments where they make up 34% of the prokaryotic biomass in Antarctic coastal waters. They have been detected in glacial ice, but primarily at the bases of glaciers, where they find the nutrients that they need. They also have been found in the sediments of subglacial Lake Whillans and in small numbers in subglacial Lake Vostok. So far, those growing in basal ice and in subglacial lakes have been methanogens, combining carbon dioxide with ammonia to produce methane, and in doing so gain the energy they need for their metabolic needs. None of these are pathogenic, as far as we know.

EUKARYA

In addition to bacteria, many types of fungi, including those that are pathogenic on animals and plants, survive freezing and thawing. Several types of protists also survive freezing and thawing, although most do so at lower levels than bacteria and viruses. This includes *Giardia lamblia*, which lives in cold water and infects several mammal species, including humans. Species of *Plasmodium*, one of which is the causative agent of malaria, also survive to an extent when frozen and thawed. Several species of protists were found in the accretion ice from subglacial Lake Vostok, including some that are pathogenic on animals, and a few that are potentially human pathogens. One was a species of the amoeba, *Naeglaria*, which also is known to survive freezing and thawing, especially when in the cyst form. They are found in cold and warm water, and if they gain access to human nasal cavities they can begin growing, eventually leading to a fatal infection of brain tissue. Some fatal cases of *Naeglaria* infection have been contracted while swimming in ponds and lakes, and from tap water.

Fungi are eukaryotic microorganisms composed of filaments called hyphae that derive their energy and nourishment from dead or living organisms. They generally reproduce by the production of large numbers of haploid and/or diploid spores. You probably breathe in 50 or so fungal spores with every breath you take! Many of them might have blown in from the poles. Fungi grow best in dark, moist places, but can be found wherever organic matter is present. Most fungi are saprobes (nutrients obtained from dead organic sources), but some are parasitic. Most fungi require oxygen, but some yeasts also can grow in the absence of oxygen. Most fungi survive freezing and thawing. This is why if your refrigerator or freezer fail, and you only notice it weeks later, fungi will be found growing everywhere. Most are very hardy organisms, and many form spores that are readily transported by winds worldwide. Therefore, they can melt out of a glacier in a polar region, and can be blown into temperate or tropical zones, or vice versa. Large number of species are found in polar regions.

We have found many fungi in both Greenland and Antarctic ice, as well as in the accretion ice from subglacial Lake Vostok. Fungi are ubiquitous. They are major recyclers of nutrients, but it is not their only important function. Fungi are important for many reasons. For example, many antibiotics are produced by fungi, as well as certain drugs (e.g., cortisone and cyclosporine). Penicillin and griseofulvin are but two examples of important antibiotics produced by fungi. Imagine a world without penicillin. Almost all of us have taken this antibiotic (or its relatives and derivatives) at one time or another in our lives to combat a variety of infections. The bread, wine, and beer industry are absolutely dependent upon yeasts, which are nonfilamentous fungi, for the fermentation of sugar to ethanol and carbon dioxide. Fungi are used in the production of some cheeses (bleu cheese) and soy sauce. Fungi also have been important in the survival of humans, and in human civilization. Bread, beer, wine, tofu, and cheese are foods that require the use of fungi in their production. The fungi help to retard spoilage and improve nutrition either by producing vitamins, or by converting specific nutrients to forms that can be better absorbed by the human digestive system, and to provide nutrition for gut microbes. Fungi are part of the flora of the guts of animals, including arthropods, birds, mammals, and others. Some fungi form beneficial associations with plants. These associations, termed mycorrhizal associations, help the plants soak up water and nutrients through their roots. If the mycorrhizal fungi are absent, the plants may die or become severely weakened.

Many fungi are pathogens. Fungi are the major cause of plant diseases. Approximately 5,000 species are pathogens of economically valuable agricultural, garden, and woody plant species (including fruit and forest tree species). The costs involved in controlling fungal diseases of food crops amount to billions of dollars annually. In addition, fungi cause diseases in animals and produce toxins (e.g., aflatoxins, ergot alkaloids, amanitin) in contaminated foodstuffs that affect humans, poultry, swine, cattle, and horses. Many infect fish, although many of the diseases that once were thought to be fungal are actually caused by protists that have some morphological similarities to fungi. Many of these are hardy and can retain viability following freezing. Humans also are not immune to fungal infections. Fungi are the cause of ringworm, athletes' foot, various systemic mycoses (e.g., valley fever, candidiasis, chromoblastomycosis, and histoplasmosis). One species, *Candida auris*, is resistant to all known drugs that are used to treat fungal infections. It is fatal in many cases. While many fungi cause superficial infections, some can become systemic and deadly. This includes species of *Aspergillus*, and related fungi, which have been found in polar ice. In fact, fungi have been among the most often found groups of eukaryotes in ice core samples.

Some protists have also been found in glacial ice and in subglacial lake ice. Slime molds, many water molds (members of a group called oomycetes), paramecia, dinoflagellates, excavates (e.g., *Giardia and Trypanosoma)*, and many others are within this group. Historically, they have been difficult to classify, primarily because some stages of their life cycle include many very different and disparate steps, and many are the result of multiple endosymbiotic events, which results in the mixing of the genomes of two or more complex genomes. Some resemble animal cells, others resemble plants and algae, some resemble fungi, and still others are unique in their forms (e.g., foraminifera, rhizaria, dinoflagellates). There are several types of slime molds, and several types of oomycetes (water molds). One of the most recognizable oomycetes is *Phytophthora infestans*, the causative agent of potato late blight disease. This disease devastated Ireland over a century ago, and is currently a major problem worldwide, and is especially troublesome in the Americas. Another species in this genus is *Phytophthora sojae*, causative agent of soybean blight. It causes millions of dollars of damage annually. These protists survive freezing and thawing under certain circumstances. Because of the conditions needed, the survival rate under environmental freezing and thawing conditions is expected to be low, but not zero.

Several protists have been found in glacial, surface lake, and subglacial Lake Vostok accretion ice. Protozoans and sporozoans are eukaryotic, usually motile, unicellular protists. They grow in a wide variety of moist habitats. Most are free living and inhabit terrestrial, marine, and freshwater environments. Many terrestrial forms are found in decaying organic matter, soil, and beach sand. Some are parasites of plants, animals, and humans. Protozoans are important in nature for two major reasons. First, they constitute a large part of zooplankton, and thus are an important link in many aquatic food chains. Second, many important diseases of humans and animals are caused by protozoans. Most protozoans obtain their carbon and energy from organic carbon sources, that is, they eat other organisms, living and dead. Many protozoans produce a resting stage called a cyst, which is a dormant form with a thick cell wall and very low metabolic activity. These cysts protect against adverse environmental conditions, including being frozen in ice, and function to facilitate transfer of the organism between hosts. Most species reproduce asexually by binary fission. Sexual reproduction in those species in which it occurs, is by conjugation or exchange of gametes between paired protozoans of complementary mating type. The genus *Heteromita*, which has been detected in polar ice and Antarctic ice fields is a member of this group. It feeds primarily on bacteria.

An important human disease caused by protozoans is giardiasis, which is caused by *Giardia lamblia*. Millions of people worldwide (including in the US) contract this diarrheal disease each year from contaminated water. These organisms are interesting in that they are a group of protists that have no mitochondria. Although it is rarely fatal, occasionally people die from the disease due to extreme dehydration. Various Trichomonads live in the vagina and urethra of women and in the prostate, and urethra of men, and are usually transmitted by sexual intercourse. Although they can survive freezing and thawing, they also need to colonize a host soon after thawing. Trypanosomes, another group of protozoa cause many human diseases, of which arguably the best known is probably *Trypanosoma brucei*, causative agent of African sleeping sickness. However, many other species exist, including *T. cruzii*, which causes Chagas disease. Charles Darwin may have contracted this disease during his expedition on HMS Beagle, while in South America. It may have eventually led to his death. The parasite causes slow deterioration of cardiac muscle, leading to increasing fatigue over years or decades, eventually progressing to heart failure. Typanosomes have been detected in subglacial Lake Vostok ice. It is unclear what they are growing on or in, but there are indications that fish, crustaceans, bivalves, cnidarians, and arthropods also are found in the lake, which may be the host organisms. Trypanosomes often are found in arthropods, so this might be their source. Amoeba also have been found in glacial and lake ice. Severe amoebic dysentery, which may be fatal in humans, is caused by *Entamoeba histolytica*, and sequences from amoeba have been detected in Lake Vostok accretion ice. Lastly, malaria and cryptosporidiosis are caused by a group of protozoa known as sporozoans. While these organisms have lower rates of survival after being frozen and thawed, some do survive. We have found sequences from several species of protists in some of the ice and subglacial ice samples, but it is unknown whether they are viable. Culturing of these organisms is made difficult by the fact that they often only grow within a specific species of host organisms. In the Vostok ice, protists were found only in the accretion ice, but not the glacial ice above the accretion ice, hinting that the protists are living in the lake, and were not transported into the lake from the glacial ice above.

HAZARDS AND DANGERS

Scientists who work with ice realize that there are known and unknown dangers. One reason that biologists study microbes in ice is to identify some of these dangers, so that they can be monitored, managed, and mitigated. This applies to studies of global climate change, sea level rise, releases of pathogens, and others. As it is now, there are many more unknown dangers in the ice. Therefore, it is difficult to guard against all of the dangers when working with ice. Some are relatively minor, but others can be life threatening. Ignoring the dangers of melting ice is not an option. When working with the ice, extraordinary caution is practiced. When we first started isolating and sequencing these organisms it was more-or-less a blind effort, not knowing what we would find. Once we had a large number of cultures and nucleic acid sequences, we could finally start looking for trends in the types of microbes represented in the ice. The first conclusion was that there were many organisms, and the diversity was higher than expected. Some of the organisms that we found had us scratching our heads, and we had to think deeply about why they were there. The second conclusion was that many of the organisms that are found in the ice are floating in the air that we breathe every day. While this might mean that the dangers are minimal, this is not necessarily true for people with weakened immune systems.

Some microbes that were entrapped in ice that was hundreds of thousands of years old closely resembled contemporary microbes, while others were completely unknown. We still have not identified more than half of them! No one knows at this point whether these simply have never been studied, they represent extinct species, or are only found in ice. All we know is that their sequences are unlike any that have been deposited in national and international sequence databases. Once we started to isolate many microbes we could not identify, there began a nagging fear in the backs of our minds. If any of these were pathogens, particularly unknown or virulent pathogens, then we would have to be extremely careful working with them. If they were pathogens unknown to science and medicine, then there might not be treatments for those infected. On the other hand, the same types of microbes have been melting from glaciers for millions of years. If they were deadly, wouldn't we have been aware of this sooner? Perhaps, but this was only a partial consolation. For those who live near glaciers and use the meltwater, the preserved pathogens present a continuous, albeit small, risk. The dangers encased in ice are real, and with the increases in melting of the ice worldwide, the risks from the release of pathogenic microbes also are increasing. Just to be safe, the work in our labs with potentially viable organisms is performed in a biosafety hood within an isolated room. The room is regularly decontaminated and bathed in ultraviolet light. All of this protects the personnel in the lab, and prevents release of the organisms from the lab.

Ice cores have been drilled in many places around the world, but the highest densities of cores and the deepest cores are from Greenland and Antarctica. The deepest cores from Greenland are from GISP2 (for Greenland Ice Sheet Project, site 2) and GRIP (European Greenland Ice Sheet Project). Both are more than 3,000 m (10,000 ft) deep. The deepest ice was deposited from 100,000 to 200,000 years ago. The Vostok, Antarctica core is over 3,700 m (12,000 ft) deep. The ice at 3,538 m (11,600 ft, the deepest of the basal ice) might be more than one million years old. There is older ice on the Earth, some is several million years old. Some permafrost is even older than this, up to 25 million years old. Travel to these sites often is hazardous. Construction of the camps and extraction of the ice is challenging and expensive. The first step in the process of obtaining and studying ice cores is to travel to the drilling site. This often takes years of preparation, millions in funding, and sometimes involves travel to relatively inaccessible, dangerous, and extremely cold places. The logistics of getting people, equipment, and supplies to the site is time-consuming, difficult, dangerous, and expensive. The first decisions involve where to drill, how deep to drill, what ice sections are important, and how much ice is needed. Not surprisingly, this is a lengthy, and sometimes difficult and competitive process. Initial tests are done at several potential drill sites. Many meetings and many presentations are made. In the end, the decisions are based on the best science, having the proper teams assembled, finding appropriate funding, and choice of locations that logistically are possible. Being logistically possible does not mean easy, or that all dangers are eliminated. The equipment is powerful, the locations are remote, and the temperatures are hazardous. Lifelines consist of snowcats, sleds, portable housing, sturdy tents, warm clothing, good boots, food, heating sources, helicopters, and airplanes. At -20°C to -80°C, sometimes with high winds, reliable shipments of food and fuel are absolutely vital. Travel to Greenland and Antarctica is somewhat risky, but travel to the Andes, the Himalayas, the Rockies, or the Alps also presents many dangers. Occasionally, the workers and researchers have been injured or become ill, which has necessitated emergency evacuation. A few have died from their injuries or illnesses, but fortunately these have been rare.

Once the personnel, drilling apparatus, and housing issues are settled, drilling begins. Drilling of shallow cores can be relatively simple, and so-called "dry drilling," that is drilling without the use of fluids to smooth the drilling process, can be utilized. When drilling involves reaching to hundreds and thousands of meters deep, other issues come into play. One is time, and the working seasons in the Arctic and Antarctic are short, usually just a few months. It takes more time to drill each core section as the depth increases, because the drill bit has to be lowered to drill, and then raised to remove the core samples, each of which is from 1–3 m (3–10 ft) in length. Another factor is temperature. As depth increases, so does temperature, so that near the bases of the glaciers, the temperature often is close to 0°C. The drill causes friction and heat, and the ice can melt. So, drilling fluids must be used to keep the borehole from refreezing in order to avoid having to repeat drilling the borehole. Another major issue is pressure. The deeper one goes into the ice, the greater the pressure. The drills can become wedged in the borehole. Also, at certain depths, there can be fragile ice, known as "brittle ice." When the pressure is released by drilling and bringing the core to the surface, the pressure change can cause the core to crack, split, and sometimes explode. In fact, simply tapping on the ice core can cause it to fracture into many pieces. When these core sections are removed from the drill, usually they are handled very carefully, and are left undisturbed on site and below freezing for weeks or months to allow them to "relax." After this, they can be manipulated and tested without the danger of cracking or exploding. Utilization of drilling fluids (diesel fuel, ethylene glycol, hot water, or others) alleviates some of the problems with pressure, such that less cracking and splitting occur, primarily because the release of pressure is slower when fluids surround the ice core sections. But, it also introduces contaminants onto the outer surface of the ice core sections, and to the bore hole. Once the ice is removed from the site, it is shipped frozen to storage and research facilities. Even after transport to the labs, there still are dangers. There are potentially viable pathogens in the ice.

Most often, drilling and research crews spend the summer months excavating core sections and assaying the contents. In many locales, the temperature rarely is above 0°C. On a sunny day, with no wind, it actually can feel warm, and some work in short sleeves! However, storms can quickly move in and so the workers must be aware of their situations at all times. While most are on site for only a few weeks or months, some workers stay at the sites year round. For example, for more than a few decades, a small group of workers manned the Vostok station continuously, in order to maintain the equipment in the extreme conditions. In the winter, no sun shines for months, and the temperatures range from -40°C to -70°C (-40°F to -94°F). The coldest temperature ever recorded on Earth was -89°C (-128°F) at Vostok station in 1983. Even going outside for a few minutes in the winter at Vostok can be dangerous. They must be absolutely certain that before winter sets in, they have all of the food and fuel that they need to survive through the winter. Supply vehicles and airplanes cannot be used at these extremely low temperatures, because hydraulic lines freeze at these temperatures. So, there are no supply lines possible in the winter. Many people literally risked their lives to extract these ice cores. Many continue to work in these conditions in order to study this important scientific resource.

SOURCES AND ADDITIONAL READINGS

Castello, J.D., and S.O. Rogers. 2005. *Life in Ancient Ice*. Princeton, NJ: Princeton University Press.

Christner, B.C., J.C. Priscu, A.M. Achberger, C. Barbante, S.P. Carter, K. Christianson, A.B. Michaud, J.A. Mikucki, A.C. Mitchell, M.L. Skidmore, T J. Vick-Majors, and the WISSARD Science Team. 2014. A microbial ecosystem beneath the West Antarctic ice sheet. *Nature* 512: 310-313.

D'Elia, T., R. Veerapaneni, and S.O. Rogers. 2008. Isolation of microbes from Lake Vostok accretion ice. *Appl. Environ. Microbiol.* 74: 4962-4965.

D'Elia, T, R. Veerapaneni, V. Theraisnathan, and S.O. Roger.s 2009. Isolation of fungi from Lake Vostok accretion ice. *Mycologia* 101: 751-763.

Ma, L., C. Catranis, W.T. Starmer, and S.O. Rogers. 1999. Revival and characterization of fungi from ancient polar ice. Mycologist 13:70-73.

Ma, L., S.O. Rogers, C. Catranis, and W.T. Starmer. 2000. Detection and characterization of ancient fungi entrapped in glacial ice. *Mycologia* 92: 286-295.

Nilsson, M., and I. Renberg. 1990. Viable endospores of *Themoactinomyces vulgaris* in lake sediments as indicators of agricultural history. *Appl. Environ. Microbiol.* 56: 2025-2028.

Rivkina, E., J. Laurinavichyus, and D. A. Gilichinsky. 2005. Microbial life below the freezing point within permafrost. In: Castello, J D. and S. O. Rogers (eds.) *Life in Ancient Ice*. Princeton NJ: Princeton University Press, pp. 106-117.

Rogers, S.O, W.T. Starmer, and J.D. Castello. 2004. Recycling of pathogenic microbes through survival in ice. *Med. Hypoth.* 63: 773-777.

Rogers, S.O., Y.M. Shtarkman, Z.A. Koçer, R. Edgar, R. Veerapaneni, and T. D'Elia. 2013. Ecology of subglacial Lake Vostok (Antarctica), based on metagenomic/meta transcriptomic analyses of accretion ice. *Biology* 2: 629-650.

Rogers, S.O., V. Theraisnathan, L.-J. Ma, Y. Zhao, G. Zhang, S.-G. Shin, J.D. Castello, and W.T. Starmer. 2004. Comparisons of protocols to decontaminate environmental ice samples for biological and molecular examinations. *Appl. Environ. Microbiol.* 70: 2540-44.

Shtarkman Y.M., Z.A. Koçer, R. Edgar R, R.S. Veerapaneni, T. D'Elia, P.F. Morris, and S.O. Rogers. 2013. Subglacial Lake Vostok (Antarctica) accretion ice contains a diverse set of sequences from aquatic, marine and sediment-inhabiting Bacteria and Eukarya. PLoS ONE 8(7): e67221. doi:10.1371/journal.pone.0067221.

LINKS

http://www.antarcticglaciers.org/glaciers-and-climate/ice-cores/ice-core-basics/
https://www.researchgate.net/figure/Map-of-Greenland-showing-the-position-of-ice-core-records-red-and-meteorological_fig1_286417720

17 Everything Old Is New Again, the Ultimate in Recycling

"I was so much older then, I'm younger than that now"

Bob Dylan

AN UNFORTUNATE ACCIDENT

On the night of August 2, 1947, a British South American Airways airplane with six passengers and five crew was on final approach into the Santiago, Chile airport, having come from Buenos Aires, Argentina. The pilot was in radio contact with the airport tower. Everything seemed normal, except they reported that they were flying through clouds. They were over the snow and ice-covered Andes Mountains, descending into the airport, with only about 4 minutes left before landing. Then, a mysterious Morse code radio message came through from the radio operator on the plane that was undecipherable. No one could understand the message, but the Morse code message may have indicated trouble on the flight. The plane was possibly in a stall. They heard nothing more from the radio operator. The plane and passengers never arrived at the airport. The search went on for weeks. Investigators examined all of the possible locations for the crash, but they failed to find the plane. Half a century later, a hiker on the side of an Andean mountain found a wheel and tire sticking up from the end of a glacier. Although hundreds of hikers had passed near the same spot over the past 50 years, no one reported seeing the wheel. After reporting the find, others came and found parts of the engines and propellers, including a tire that still was inflated. Part of the fuselage also was located nearby. It was the missing plane. It was found coming out of the end of a glacier that starts much higher on the mountain. The time it takes for an object entrapped near the top of the glacier to reach the bottom face of the glacier (the cycling time of the glacier) is about 50 years. This is the same whether the object is an airplane or a microbe. When the plane hit the mountain, it must have triggered an avalanche that buried it near the top of the glacier. A fresh coat of snow hid all traces of the impact. Over the years, the plane and everything inside was compacted in the ice and moved as the glacier moved, slowly traveling down the mountain trapped within the river of ice. Overall, it remained much as it was just after the crash. In fact, crash investigators could still determine that both

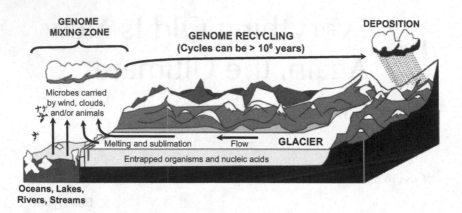

FIGURE 17.1 Diagram of the concept of genome recycling. Snow and ice, as well as microbes, are deposited on the upper portion of the glacier (right), which becomes compacted by additional snow and ice. The glacier carries the microbes as it flows downhill, which eventually are released through surface sublimation or melting at the end of the glacier. If the microbes are released in a suitable environment, are carried to a suitable environment, or come in contact with a host organism, they may be capable of growth and spread. Some of the resulting microbes may be transported onto another (or the same) glacier to begin another cycle.

engines were running at high speed at the time of the crash, possibly indicating that they were in a stall, and the pilot was trying to regain control of the plane. The people, too, remained much as they were at the time of the crash. Unfortunately, all 11 perished in the crash. While people cannot emerge alive from being frozen in a glacier, many microbes can. As they melt from the glaciers, they carry on much as they did before they were frozen in the glacier. This phenomenon, which we have termed "genome recycling," has some interesting implications (Fig. 17.1).

STOPS AND STARTS OF LIFE

Environmental ice provides a means for preservation of a large number of microbes and their genomes over time. However, the entire process of genome recycling is dependent on the revival and establishment of organisms once they emerge from the ice. They must either land, migrate, or be transported to a suitable location. Also, they must successfully interact with other organisms in the vicinity, as well as the extant populations of their own species. Therefore, they must be present in sufficient numbers and propagate, in order for their genomes to survive. The existence of accessible habitats, transport mechanisms, sufficient numbers of individuals, and alleles that confer adaptability to the extant environmental conditions are necessities. Local and global weather patterns, as well as substrates for specific organisms, determine the probability of growth, migration, deposition, release, and gene flow (i.e., entry of their genomes into existing populations). Success of the ancient genomes depends on their compatibilities with current populations, host systems, and the overall (as well as specific) environmental conditions.

Viable microorganisms are spread globally by wind, riding on dust, water droplets, snow, and ice pellets. Microbes settle out along the way, based on their densities (or the densities of the particles to which they are attached), wind velocities and directions, and whether the water or ice precipitates, or whether the wind is flowing over land or water. When they become incorporated into ice, those that survive freezing and thawing may persist ephemerally, or for extremely long periods of time. If they survive thawing of the ice, they have the potential to interact with modern populations of the same species or with their host organisms (if they are pathogens, parasites, or commensals). Since their metabolism and reproduction levels while they are frozen were extremely slow, they may be genetically different from the populations that were unfrozen and active during the same time period. This is because replication and cell division always lead to some amount of genetic mutation. The effect of this is that some genes and versions of genes (alleles) essentially were stored frozen while in the ice, while alleles in the dividing population underwent constant change. Upon melting, the previously entrapped microbes enter the modern populations. Therefore, some of the preserved ancient genes are recycled into the existing populations.

GLACIAL CYCLES

Glaciers, ice domes, and ice fields entrap and release huge numbers of microbes, sometimes for more than 500,000 years. Since the last glacial maximum, from 16,000 to 31,000 years ago, glaciers have been releasing enormous numbers of ancient microorganisms. Over the past 1,000 years, the rate of glacier retreat appears to have increased. More recent accelerations in glacial recession have been recorded, a major one during the 19th century, and a further acceleration during the 20th and 21st centuries. Almost all of the ice on Mt. Kilimanjaro, Africa is gone, the majority of which has melted in just the past 100 years. All of the ice is gone from Mt. Kirinyaga (formerly Mt. Kenya), Africa. Many glaciers around the world are disappearing, kilometers at a time. Glacier National Park soon will need a name change. Photographic and written records of glaciers in the Alps document the recession over the past three centuries. These records are startling, and should cause anyone to sit up and take notice. This represents the melting of enormous amounts of ice, with concomitant huge releases of entrapped microbes.

Glaciers constantly flow. They flow much like rivers, only slower. Some flow so slowly that it is difficult to discern their movement. Others flow so fast that you can watch them move, and you can hear them crack and pop. Snow deposited on the upper parts of glaciers is compacted, and then the glacial ice flows slowly down inclines. Thus, the oldest ice generally is at the bottom end of the glacier, where it interacts with the underlying rock, and then deposits boulders, stones, gravel, sand, water, dissolved minerals, microbes, and other inclusions (including people). Ice melting today and entering the streams, lakes, and oceans may be centuries or millennia old, or older. The microorganisms melting out of the glacier have been absent from the active environment for very long periods of time. In that time, individuals within populations of the same species on the land, in the atmosphere, and in oceans, lakes, and streams have undergone countless replication and cell division cycles. During each of these cycles mutations occur. Therefore, the organisms melting out of the ice and the individuals of the same species out in the active environment may have different alleles (gene versions) and allelic combinations. These genetic differences may not be useful in the modern environment into which they are released, which might cause them to die. On the other hand, some genetic changes may lead to large advantages for the microbes once they are released into the modern environment. Some alleles may confer a selective advantage to some of the ancient microbes, such that they overwhelm competing organisms. One example of this can occur with pathogenic organisms. Normally, hosts and host populations become immune or resistant to particular strains of pathogens. Think of yourself. You were infected with, or vaccinated against, certain viruses once (or twice, if a booster vaccination was required). You never contracted the disease after that. Why not? You had acquired immunity to this virus. However, if this virus could be sequestered for one or more human generations, it could reemerge and infect a population of humans that had never acquired immunity (unless they were vaccinated). This is a concern with smallpox virus reemergence. Very few have been inoculated for this virus, and if it were to reemerge, most of the human population worldwide would be susceptible to it.

Several research groups have demonstrated that approximately 10^3 to 10^7 viable microbes are present in each liter of environmental ice meltwater. Currently, approximately 500 billion tons of ice is melting from glaciers, ice fields, and ice domes that is not being replaced. This means that in an average year at least 10^{18} to 10^{22} viable microbes (including fungi, bacteria, and viruses) are released from environmental ice. This is equivalent to 10^4 to 10^8 tons of microbial biomass. If spread evenly over the surface of the Earth, this would be approximately 10^3 to 10^7 propagules (cells or virus particles) per species per square meter. Tens of thousands of species may be represented. However, the past two decades have been anything but average. The Earth is warmer, the polar regions are warming faster than any other locales, and there has been a noticeable increase in the rate of melting of ice worldwide. Therefore, the actual releases of microbes may be much greater. Microbes are found in clouds, fog, rain, snow, and elsewhere in the atmosphere indicating that transportation mechanisms are readily available. Animals, oceans, streams, wind, and other forms of transportation also are common. Depending on wind currents, season, and precipitation, propagules may be distributed worldwide, thus increasing the chances of landing on a suitable substrate, including plants or animals. This represents a migrating population large enough to become established and to interact genetically with contemporary populations virtually anywhere on Earth. Potentially, local outbreaks, epidemics, and pandemics may have been caused by this route, but because no one is monitoring microbes melting from ice, it is unknown whether or not this has occurred.

THE 1918 INFLUENZA A PANDEMIC

Influenza A virus strains come and go. Some are relatively mild, and some are much more debilitating. All are capable of causing deaths. The influenza strain that caused the pandemic during World War I, and was at its peak in 1918, killed at least 50 million people worldwide. As stated in a previous chapter, although it is often called the "Spanish" flu, it probably already had its start in China in 1917, and the genomic pieces of this virus probably had their beginnings earlier than 1917. While it clearly was influenza A, this one was different. It killed not only infants, elderly, and weaker individuals, but it killed healthy young people. It was the only time in modern history that populations, including the US population, decreased. If this strain could be entrapped in the ice and reemerge, it would have the potential to create a deadly epidemic or pandemic. What if the 1918 flu strain originated from ice melt? We don't know if this happened, but now we believe it is possible. We have found that influenza A viruses survive freezing and thawing in Siberian lakes, and others have also found viable influenza A in ice, cold water, and mud within ponds and lakes. Therefore, influenza A virus and other viruses are preserved in ice, possibly for long periods of time, and they are capable of surviving when they melt from the ice at a later time. Studies of virus survival under cold and freezing conditions, indicate that the majority of viruses are capable of surviving freezing and thawing, and this includes many pathogenic species and strains.

PATHOGEN CYCLING

In the past, it had been noted that strains of pathogenic organisms (e.g., influenza viruses, caliciviruses, polioviruses, and tobamoviruses) disappeared for years, decades, centuries, or longer, and then reappeared to infect again. For example, there have been reports of the appearance of genetically identical (or very similar) influenza A (an RNA virus) strains separated by decades of absence. The length of time for their disappearances and emergences have ranged from a few years to over 30 years. This last number approximates a human generation, which is also about the amount of time it takes for a population to lose sufficient immunity to a particular pathogen for the population to become susceptible to that pathogen again. The appearances, disappearances, and reappearances of caliciviruses also could be explained by polar ice entrapment and later release by melting. An example described in an earlier chapter outlined how a strain of calicivirus in New Jersey swine disappeared for two dozen years, only to turn up again in whales in Alaska, and in sealions in California. An explanation based on the movement of infected animals seems implausible. However, an explanation based on the virus being deposited in ice by an infected animal (or being transported by wind,

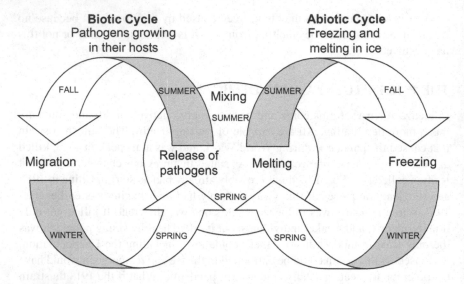

FIGURE 17.2 Biotic and abiotic cycles involved with pathogenic organisms that can survive freezing and thawing. In the biotic cycle, host organisms carrying a pathogen release the pathogens into the environment (e.g., by coughing, sneezing, defecating, urinating) into a body of water (e.g., migratory birds during the Spring and Summer), but also drink the water, which usually contains a mixture of pathogens. The hosts may migrate, carrying the mixture of pathogens with them. The abiotic cycle includes the freezing of the pathogens in the water, thus preserving them during the Fall and Winter months. As the ice thaws during the Spring, the pathogens melt out of the ice and are picked up by susceptible hosts. The hosts also deposit new mixtures of pathogens, which then are frozen in the ice during the Fall and Winter. The ponds, lakes, rivers, and streams become mixing vessels, containing newly arrived mixtures of pathogens, as well as the previously deposited mixture of pathogens that were frozen in the ice.

waves, currents, or precipitation), and then thawing out much later is more plausible. Upon melting from the ice, the pathogens can find hosts that lack immunity or resistance to the pathogen, and thus they are able to easily infect individuals in the population (Fig. 17.2). Previous studies and health surveillance efforts have ignored these reservoirs. Only recently have investigators begun to understand the process of release of viable microbes from glaciers and other environmental ice. Study of the abiotic sources of pathogens is important to our understanding of epidemics and pandemics that are caused by these pathogens.

MUTATION RATES, GENE FLOW, AND POPULATIONS

Genome recycling also has an effect on another genetic phenomenon, called gene flow. Gene flow is the process by which one version of a gene, carried in the genome within an individual or migrating population, is introduced and becomes fixed in a population having a different version of the same gene. In the case of frozen organisms, this would be temporal (time-based) gene flow. Depending on the number of organisms coming out of the glacier into the new population, the population of cells emerging from the ice could have from no effect to a huge effect on the existing population. If the contemporary population is large, and the thawed population is small, gene flow might be zero, especially if the ancient alleles have no selective advantage over the newer alleles. However, if the contemporary population is small and the thawing population is large, or if the thawing population has advantageous alleles (or allelic combinations), the ancient genotype could overtake and replace the newer genotype. The likely result would be somewhere between these two extremes. This might be something to consider, given the accelerating rate of glacial melting in the world. In the past few decades and centuries, much more mixing of ancient and modern genotypes has been occurring. Some mixing might be beneficial to humans, but it is equally likely that some mixing might be detrimental. In fact, this might be more hazardous to you than being attacked by a foreign enemy. However, it will be equally hazardous to your enemy.

One of the reasons we began to study life in ancient ice was to find a way to measure mutation rates in these organisms over a long time scale. In essence, we thought that it would be possible to observe evolution in action. We now know that this is more difficult that we had originally surmised. The first problem is one of population. At a specific depth of ice, we can isolate, sequence, and/or otherwise identify an individual fungus, bacterium, or virus (although sometimes this is not a trivial matter). At another depth we might find another cell of the same species. However, their genotypes are usually somewhat different. If they are directly related to one another by genealogy, then mutation rates can be calculated, as long as the ice core sections have been dated accurately. However, if they are from subpopulations of the original population, we may just be measuring the genetic variation within the original population, which is not the same thing as calculating the mutation rate.

There is another problem. The two isolated microbes could have originated from two geographically and/or temporally isolated populations. In this case again, we cannot calculate mutation rates based on the differences between the two because the two may have been genetically isolated from one another long before their deposition in the ice. There are other problems, as well, the primary one of which became an unexpected, but an interesting part of our research. Because the organisms are inactive (or nearly so) during their encasement within the glacier, they are not replicating their DNA (or RNA in the case of some viruses). Therefore, the rates of mutation are nearly zero. This is what we were counting on originally, and it appears to be so. However, when the organism reenters the modern world after it melts out of the glacier, it is entering a world different from the one it left when it was frozen in the ice. Also, it is entering a population of its own kind that has continued to grow and divide during the years that the organism was frozen in the glacier. Therefore, the population that continued to grow and divide has accumulated new mutations (and perhaps genes) that are not present in the "Rip van Winkle" microbe. If there are enough of these old organisms that enter the extant population of the same kind of organism, they can effectively change the proportion of variant genes and alleles in that new population. Even if there are very few of these ancient organisms, if their genes confer selective advantages for survival, they and their genes may become firmly established in the population. Therefore, they may be able to change the allele proportions in the population. This is the essence of genome recycling, which is actually a form of temporal gene flow. However, this means that mutation rates cannot easily be determined by examining genomes from different depths in the ice.

Genome recycling causes a problem for scientists trying to measure mutation rates, because there are mixtures of new and old genotypes circulating through the population. Thus, when the mutation rates are measured, a lower than expected value might arise for particular genes or particular parts of the microbial population. This is because the lower mutation rates in the organisms that were frozen are lowering the result of calculating the actual mutation rates of the non-frozen and active populations. There has been a mystery regarding mutation rates in bacteria and viruses for several decades. Specifically, they appear to have mutation rates well below what their genomes could handle. Could genome recycling be the cause? It is a possibility. Microbes have been constantly recycling old genotypes into resident populations, sometimes causing the old genes to replace more recent versions. In some cases, the older genes can become completely incorporated into the microbial populations, thus causing the apparent absence of mutation over long periods of time. This could be a benefit to various species, in that they can essentially keep trying new versions of a gene, while keeping various versions in cold storage that are brought back from time to time, just in case they are more advantageous to that species later. This might be very important in an evolutionary sense, because it is another way in which allelic diversity can be maintained, and multigene families can evolve. The two versions of the gene (the older version and the newer version) may come together on the same chromosome via crossing-over, such that a single organism would contain both versions of that gene. This would allow one version to change more rapidly, eventually developing additional functions, and establishing a multigene family.

Another advantage to these microbes is that a portion of the population can be saved by freezing during environmentally stressful events. If there is a worldwide heat-wave, fires, asteroid impacts, drought, volcanic eruptions, ice ages, or other environmental catastrophes, species may be saved by being frozen in glacial ice. The extant populations may become extinct, but as long as some members are frozen in glaciers, the species may survive. Pathogens would have a clear advantage in this respect. If a host population gained resistance or immunity to a pathogen, then the extant population of the pathogen might go extinct. But, over time hosts lose resistance or immunity in the absence of the pathogen. Thus, the frozen population need only remain frozen for decades or centuries. Then, when the individual pathogens melt from the glaciers, they encounter immunologically naïve hosts that can be easily infected. In plants, animals, bacteria, protists, and viruses, the outcome is similar, although the process to acquire immunity differs. The pathogen cannot reproduce itself in those populations that have become resistant or immune to it. However, if it can be frozen in ice and wait for a host generation or so, the host organisms with resistance and immunity will die, and their offspring may not have acquired immunity to the pathogens. Thus, once the pathogens melt out of the ice, they may find a completely susceptible population of hosts. This could happen with the 1918 influenza strain. Almost all of the people who were exposed to that strain are now dead. Thus, the current population may be susceptible to this deadly strain of virus. If the 1918 influenza reappeared today, tens-of-millions (or more) could die. Other viruses appear to exhibit this same strategy. Certain strains of virus disappear for decades, and then reemerge virtually unchanged from when they disappeared. They can then infect an entirely susceptible population of host organisms and cause significant morbidity and mortality.

IMPLICATIONS

One of the predictions of genome recycling is that at each point in the glacier there will be a mixture of genomes, meaning that there is no clear separation between ancient and modern genomes, but that there is continual mixing. This is exactly what we have found in our analyses of viruses, bacteria, and fungi that we have studied in glacial ice. This has a direct effect on allele frequencies, and effectively lengthens the life cycle of the organism. In fact, if one considers entrapment in the glaciers as a part of the life cycles of these microbes, some of them may have some of the longest life cycles of all organisms. Genome recycling may be a mechanism that enables microbes to survive harmful events on Earth. Although genome recycling has yet to be explicitly proven, the evidence is mounting. The results from many different types of studies are consistent with the predictions of genome recycling. The major indication that genome recycling exists is that there is a homogenization of genomes due to gene flow between ancient and modern conspecific organisms. The evidence for long-term survival in ice has been demonstrated in a broad range of studies from dozens of researchers throughout the world. Therefore, that part of the puzzle is becoming clear. The fact that microbes are distributed worldwide in the atmosphere is well established. The numbers of microbes being released from glaciers has been calculated. The homogenization of ancient and modern genotypes has been indicated in several studies. The final part of the cycle to solve is to observe genome recycling in action. Someone has to document microbes flowing from a glacier and then observe their alleles entering and becoming fixed in an extant population. This probably will occur and will be documented eventually. On the other hand, it probably is occurring daily, and has been occurring for millions (or even billions) of years, although it has not been observed directly and explicitly. If the 1918 influenza reappears, or if smallpox returns, certainly those would be notable, and they would be strong indicators of genome recycling through preservation and release in environmental ice. The current increase in melting rates of ice worldwide will put this hypothesis to the test. We only hope that the proof will not be a global and deadly pandemic.

SOURCES AND ADDITIONAL READINGS

Rogers, S. O. 2017. *Integrated Molecular Evolution*, 2nd ed. Boca Raton, FL: Taylor & Francis Group.

Rogers, SO., and M.A.M. Rogers. 1999. *Gene Flow in Fungi*. In: Worrall, J. (ed.), Structure and Dynamics of Fungal Populations, Kluwer Academic Publishers, Dordrecht, the Netherlands, pp. 97-121.

Rogers, S.O, W.T. Starmer, and J.D. Castello. 2004. Recycling of pathogenic microbes through survival in ice. *Med. Hypoth.* 63: 773-777.

Shoham, D., A. Jahangir, S. Ruenphet, and K. Takehara. 2012. Persistence of avian influenza viruses in various artificially frozen environmental water types. Influenza Res. Treat. dx.doi.org/10.1155/2012/912326.

Smith, A.W., D.E. Skilling, J.D. Castello, and S.O. Rogers. 2004. Ice as a reservoir for pathogenic animal viruses. *Med. Hypoth.*, 63: 560-566.

Zhang, G., D. Shoham, D. Gilichinsky, S. Davydov, J.D. Castello, and S.O. Rogers. 2006. Evidence for influenza A virus RNA in Siberian lake ice. *J. Virol.*, 80: 12229-12235.

LINKS

https://en.wikipedia.org/wiki/1947_BSAA_Avro_Lancastrian_Star_Dust_accident
https://en.wikipedia.org/wiki/Glacial_motion
https://en.wikipedia.org/wiki/Glacier
https://en.wikipedia.org/wiki/Spanish_flu

18 Astrobiology—Out of This World

"Somewhere, something incredible is waiting to be known"

Carl Sagan

PROBING THE SOLAR SYSTEM

People have been looking up at the stars, planets, meteors, and comets for thousands of years, and many have speculated on what is out there. Stories have been written, and movies have been produced, presenting different versions of extraterrestrial life forms. But, there have also been scientific studies regarding life on Earth and predictions of where there might be Earth-like organisms on other planets and moons. In 1976 and 1977, the Viking I and Viking II, respectively, spacecraft landed on Mars. Among the tasks that they performed were tests for the signs of life. They scooped up Martian soil samples and placed them into mixtures of chemicals, one of which was radioactive. Over the course of several weeks, the mixture was tested to determine whether the radioactive chemical had been incorporated into other compounds, ostensibly indicating biological processes. When the results were radioed back to Earth there was great excitement, because the initial data indicated that the chemical had changed in a way that meant a living organism (presumably a bacterium) was actively causing the change. There was initial excitement about the results. However, as weeks and months passed, scientists determined that certain lifeless minerals (found on Earth as well as on Mars) could produce the same results as those sent back from Mars. Additionally, another possibility that could not be excluded was that microbes from Earth might have hitched a ride on the Viking spacecraft. The Earth-originating microorganisms might be the ones causing the chemical changes rather than Martian microbes. Hearts sank and careers tarnished. Given this ambiguous (and some might say disastrous) outcome and the fact that billions of dollars were spent on this grand, but failed, experiment, it is apparent that extreme care must be taken in designing experiments to find life in unusual places. This is just as true for Earth-based studies.

Recent searches for life on Mars have focused on finding signs of water. Analyses of pictures from Mars "Rover," "Spirit," and "Opportunity," as well as satellites circling the planet, indicate that Mars once had large quantities of water, although it is too early to determine whether there is, or was, life on Mars. There are indications that pools of water exist underground, and much of the

189

polar ice caps consists of water ice overlaid with carbon dioxide ice, but whether they hold life is unknown. In 2019, readings from several Mars probes measured seasonal changes in the amount of methane in several locations on Mars. Methane increased during the Martian Summer and declined in the Winter. Generation of methane is often directly caused by methanogenic microbes, and therefore, this is an indicator that there might be living organisms on Mars. However, other measurements for methane failed to confirm these findings. All of this reminds us that it is vital to ask the right questions and apply the most appropriate methods for detecting organisms and biological processes. The important questions to ask are: 1. What organisms and molecules will be sought? 2. Where will they be located? 3. What is the likelihood of detecting the organisms and molecules, and how concentrated/dilute will they be? 4. What are the preferred methods for accurate detection? 5. What level of identification is sought (e.g., life in general, family-level, genus-level, species-level)? 6. Are the chosen methods sensitive and specific enough to detect and identify the organisms of interest? 7. How large of a sample is needed? 8. How will confounders, such as the introduction of contaminants, be eliminated from the assays? 9. How will the data be collected, analyzed, and interpreted? 10. Will samples be returned to Earth?

SEEKING SIGNS OF LIFE

Usually, the best indicators of life would be finding cells, metabolic activity, and/ or organic compounds. However, many organic compounds can be formed abiotically. For example, ten of the amino acids used by organisms on Earth can be formed without any biological intervention, and they also have been found in meteorites. Is it reasonable to expect to find signs of life elsewhere in the Solar System? Recently, scientists have been thinking that the answer is probably yes, although a few decades ago, fewer would have given that answer. The same organisms (or at least parts of them) that we find on Earth will likely be found on other bodies in the Solar System. This is true for two reasons. First, large and small bodies and particles (comets, asteroids, meteorites, dust, etc.) strike the Earth every day. They have spent long periods of time scouring space and collecting bits of debris, some of which may be microorganisms, and parts of larger organisms. When the asteroids, comets, and meteorites strike Earth they sometimes send huge clouds of debris into the atmosphere, some of which continues into space, and some that lands elsewhere in the Solar System. Thus, some organisms from Earth have likely landed on other planets and moons, just as pieces of Mars and other planets, moons, and other bodies have landed on Earth. Small fragments from the Moon and Mars have been discovered in glacial ice on Earth. Secondly, we have now sent space probes from Earth to the Moon, Mercury, Venus, Mars, Jupiter, Saturn, Uranus, Neptune, Pluto, and the moons of some of these planets. Even though scientists and engineers have attempted to assure that each spacecraft is as clean as possible, there is no way to assure sterility. Thus, these spacecraft have likely deposited microbes from Earth onto other bodies within the Solar System. Therefore, Panspermia and Exobiosis (seeding of organisms from distant locations) have occurred, and mankind has participated. The amount of contamination of the other bodies has been very small, but consideration of this contamination will have to be made when attempting to find life in other places in the Solar System. However, it is unlikely that the organisms could survive in those other worlds. If the organisms somehow could survive on any of those bodies, then the contamination issue will be even greater, because they would probably reproduce to some extent. This will muddle any conclusions of finding signs for extraterrestrial life.

It was once thought that life only existed in a very narrow zone, often called the "Goldilocks Zone," which was not too hot and not too cold, not too dry, and was a certain distance from our Sun. Outside of that, nothing could live. Now, we know that this was too narrow a view, because organisms have been found kilometers deep in caves, kilometers into the atmosphere, in hot pools, in deep hot and cold oceans, and in terrestrial and sea ice. It now seems that almost everywhere scientists look, they find life. Living organisms are found in wet, dry, hot, temperate, cold, and frozen soils (permafrost). They exist high in the atmosphere to the bottom of the oceans. In recent years, more than 10,000 new microorganisms have been isolated into culture from deep below the surface, but only a few have been thoroughly characterized and identified so far. Many viable bacteria were isolated from deep sediment cores dating back to the Cretaceous Period about 100 mybp. Approximately 1,500 bacterial isolates were examined, and many of them appeared to be new to science.

From where did these microbes originate? One possibility is that bacteria were present in the sediment at the time of deposition, remained viable, and slowly evolved into strains adapted to existence in the stable, subsurface environment, eventually evolving into new species of bacteria. Therefore, one can view deep soils and rock as being evolutionary sieves, selecting microbes from populations that possess combinations of genes that allow them to survive in these seemingly harsh conditions. A major surprise was the abundance of aerobic bacteria (oxygen-breathing), as well as those that grew only at moderate temperatures, derived their energy from organic carbon sources, and could grow rapidly in the laboratory under high or low nutrient conditions. Such bacteria are not the kind that you would expect to find in a hot, low nutrient environment with low levels of oxygen. Most microbiologists thought they would find endospore-forming (endospores are highly resistant structures produced to carry the bacterium through severe environmental conditions), anaerobic (living without oxygen), heat-loving bacteria. These rarely were found, but anaerobic, heat-loving bacteria were isolated from 2.7 km (1.7 mi) below the surface under conditions of extreme heat (75°C, 167°F) and salinity in an exploratory gas well near Washington, DC. Rock-eating bacteria and archaea, that derive their energy from iron, manganese, sulfur, chromium, and other metals, have been recovered from over 3,350 m (11,000 ft) below the surface in South African gold mines. These cells are believed to be 80–160 million years old, and are so-called "living fossils." These were found at previously unimaginable concentrations of 100,000 to 1,000,000 cells per gram of rock under conditions of extreme heat. One theory is that at least one of the species of microbes found here may actually sequester and concentrate gold, eventually leading to the formation of the gold veins in these mines. Valuable microbes, indeed.

Approximately 60 tons of space dust falls to Earth every day, amounting to more than 20,000 tons (more than 40 million pounds) of material annually. Some of this material falls on glaciers and is preserved, and some of this material may contain organic molecules and extraterrestrial microbes (if they exist). To date, none have been identified as such, although a large proportion of the sequences from metagenomic and metatranscriptomic studies were not similar to any sequences in national and international databases. This amounts to millions of sequences, which remain unclassified. Several of the missions to the Moon, Mars, and other planets and their moons have looked for signs of water and life. Although life has not yet been detected, there is a great deal of water and ice within the extraterrestrial bodies that make up the Solar System, much more than thought previously. Measurements of particular chemical signatures were made in some cases, and a few samples were examined by microscopy. Indicators of metabolic processes were performed in a few cases. The results have been promising and exciting, but still equivocal.

WHERE IN THE SOLAR SYSTEM TO LOOK?

Because it is extremely expensive to send probes and people into space, and to search for life in space, a great deal of study and planning must be performed prior to any mission. Where are the locations in the Solar System where there will be the highest possibility of detecting life? There are two categories of planets and moons that might be good places to look for life. First, these would have at least some liquid water, although each of these would also have large quantities of ice. Second, planets with ice, but not necessarily any liquid water. Because ice entraps microorganisms, both living and dead, it is possible that the ice has entrapped microbes that originated elsewhere in the Solar System that then fell onto the surface of the ice, due to the gravity of the planet or moon. As such, the icy worlds in the Solar System have acted as historic biological repositories. Among these organisms may be microbes that originated on Earth.

INNER PLANETS

The two planets closest to the Sun are not likely targets for the search for extraterrestrial life. The surface of Mercury can fluctuate between 427°C (800°F) during the day to -173°C (-280°F) at night. It is dry, and has a thin atmosphere of hydrogen, helium, and oxygen. Although there might be a very small amount of ice at its north pole (which always faces away from the Sun), the possibility of finding life there is very slim. Plus, the extreme temperatures present many challenges to designing a spacecraft to study Mercury. Venus is also a very unlikely candidate to support life. It is completely enshrouded in clouds. But, these are clouds of sulfuric acid, not water vapor. They exist within a dense atmosphere that consists mainly of carbon dioxide gas. This creates a greenhouse effect, which causes the surface temperatures at the surface of the planet to reach 467°C (872°F), hot enough to melt lead. The thick atmosphere also exerts 90 times more pressure at the surface of the planet than the atmosphere on Earth. Under these conditions, life, as we know it, probably does not exist, unless it resembles the life near deep ocean hydrothermal vents on Earth. And, reaching the surface of Venus would be difficult, due to the acid conditions and the high pressures. Ten USSR (Union of Soviet Socialist Republic) probes, and one US probe, did descend onto the surface of Venus, and did send back pictures of the surface of the planet. But, the transmissions lasted only a short time (23 to 120 minutes) on the surface, and then ceased abruptly. Four initial probes also were sent by the USSR, but all were crushed by the atmospheric pressure before they reached the surface. While the surface of Venus probably does not support life, from 50 to 60 km (31 to 37 mi) high in the atmosphere, the air pressure, composition of gases, and temperature actually are Earth-like. Not only could Earth microbes survive here, but humans could live there. It has been suggested as a location for a space station for humans. This would be the only location that might have a slim chance of supporting any life.

THE MOON

The closest body to the Earth is the Moon, but it has very little ice, some trapped within volcanic rocks, and some possible subterranean pools. Contamination is another concern with the Moon. A total of 44 spacecraft from Earth have landed or crashed on the Moon, carrying all of their microbes with them. Six of those landings included astronauts who walked and/or drove on the surface of the Moon. These people and vehicles were not sterile. Not long ago, we participated in a metagenomic study of the International Space Station (ISS). The air and surfaces all had a diversity of bacteria and fungi, which were present at the time the station parts were built, and they are brought by each astronaut who visits the ISS. If organisms were to be found on the Moon, it would be difficult to determine whether they originated on the Moon, or were simply contaminants brought from the Earth. If life is subterranean, then it might be possible to distinguish these organisms from anything that was accidentally brought to the Moon.

MARS

Mars is a better candidate, both for finding extant life, as well as searching for preserved life in ice. The two polar ice caps are 70% water ice, as well as some carbon dioxide ice, and beyond the poles permafrost extends one-third of the way to the equator. Based on the fact that small meteorites that originated on Mars have been found embedded in polar ice on Earth, it is probable that pieces of rock and ice from Earth have landed on Mars. The oldest ice on Earth so far found is 8 million years old. The oldest ice on Mars is likely much older than this. Therefore, the ice on Mars may hold a much older record of microbial life than that on Earth. However, if all life is currently extinct on Mars, then the concentrations of organisms in the ice might be very low, but remnants of the extinct life might still exist in the ice and permafrost. Clearly, the first place to look for life on Mars should be in the polar ice. If there are extant organisms growing on Mars currently, then they would need water. The largest source of water on Mars is at the poles. It is estimated that if all the polar ice on Mars melted, it could cover the surface of Mars with 11 m (36 ft) of water. So, Mars has an abundance of water, but it is primarily locked up as ice. However, as mentioned in a previous chapter, when encased in ice, some organisms on Earth surround themselves with a layer of liquid water and continue to exhibit metabolic activity. Therefore, it is possible for organisms similar to those on Earth to survive on Mars. But, because of its atmosphere, they might be primarily anaerobic. As with the Moon, several Earth vehicles have orbited, landed, and crashed on Mars, each carrying a set of Earth organisms with them. The US has had 8 vehicles (including mobile robots) that successfully landed on Mars. Six other probes (one US, three USSR, one European, and one joint project Europe/Russia) have either crashed or failed to communicate once they landed. Therefore, again, if organisms were to be detected on Mars, it might be difficult to determine their origins given that 14 probes sent from Earth landed on the surface, carrying at least some contaminating organisms with them. Although Mars has two moons, Phobos and Deimos, they are both small, irregular in shape, with no atmosphere, and are probably asteroids captured by the gravity of Mars. One USSR probe was sent to land on Phobos, but the landing was never attempted because the Earth-based station lost contact with the vehicle.

CERES

Moving out from Mars, Ceres is a dwarf planet that is similar in size to some of the larger moons, and is about 1/3 the size of Pluto (another dwarf planet), but orbits within the asteroid belt. Because of its unique characteristics, it has been proposed that Ceres originated in the Kuiper belt at the far fringes of the Solar System, and therefore is very different than the asteroids around it. Pluto, Charon (one of Pluto's moons), Phoebe (one of Jupiter's moons), and several other moons also appear to have originated in the Kuiper belt. These may hold clues about the early Solar System, and the origin of life in the Solar System. Ceres is composed of approximately 75% rock and 25% ice, but is rich in carbon and carbon compounds (20% of the surface materials), and thus has some of the basic building blocks necessary for life. It has a large icy mantle, which might have been a large subterranean ocean long ago. Whether it once had life, or whether it currently supports life is unknown, but given the amount of water and the high concentrations of carbon on the surface, it is a possible location for life.

JUPITER AND ITS MOONS

Next out from the Sun is Jupiter, which has 79 known satellites, including four large moons. It is unknown whether Jupiter itself contains any habitats for living organisms, or whether any sort of life is possible there. Although it has a rock and ice core, the bulk of the planet consists of liquids and gases, primarily hydrogen (75% by mass) and helium (24%). Jupiter has a magnetic field that is fourteen times stronger than that on Earth. This protects its atmosphere from the effects of the solar wind. It also protects the inner moons, such that some of them have substantial atmospheres that are not stripped away by the solar winds. This includes the larger moons, Io, Europa, Ganymede, and Callisto, as well as some of the smaller moons. Although Jupiter has 79 known moons, the four large moons have the greatest possibility of supporting life. Io has active volcanoes, partly caused by the gravitational forces exerted on it by Jupiter. The volcanoes produce an atmosphere rich in sulfur dioxide. If there is life on Io, it might be very different than anything on Earth. However, some of the compounds ejected from the volcanoes may migrate to Europa, which orbits close to Io. This may enrich Europa, which is covered in ice, with chemicals that are necessary for some life forms.

Ganymede, the largest moon in the Solar System, is larger than the planet Mercury (about 40% of the diameter of Earth), and is an icy moon. It has a thin atmosphere of elemental oxygen (O), O_2, and O_3 (ozone), as well as elemental hydrogen (H). Its surface is composed of mainly water ice, with an average temperature below -130°C (-200°F). It appears to have a salty ocean that is covered by about 200 km (124 mi) of ice. Although the pressure at this depth would be great, it is possible that the oceans could support life. The major drawback for study of the ocean is that there is currently no technology that would be capable of drilling 200 km into the ocean. Callisto is smaller than Ganymede, and slightly smaller than Mercury. It has a thin atmosphere consisting of elemental oxygen and carbon dioxide. It is composed of mainly rock and ice, although silicates and organic compounds have been detected on its surface. It may have a subsurface water ocean 100 km (62 mi) beneath its surface, again a possible location for life.

The moon that has generated the most interest in the search for extraterrestrial life is Europa. This moon has angular rifts and ridges crisscrossing its surface, which is completely composed of water ice. This is similar to some sea ice on Earth, where large slabs of ice push on one another, forming cracks, rifts, and ridges, although on Europa, part of the distortion of the surface is caused by the gravitational forces exerted by Jupiter. The ice on Europa is 100 km (62 mi) thick, with an ocean beneath. The currents, pressures, and turbulence in the underlying ocean cause plumes of water and ice to erupt from some of the cracks, which indicates pressure releases from the ocean below. If there are microbes living in the ocean, some of those would be pushed to the surface in these plumes, and therefore they should exist frozen in the ice on the surface. Therefore, there might not be a need to design the technology to drill through the 100 km (62 mi) of ice to reach the ocean. Signs of life could be present at the surface. Europa would be a good place to start looking for extraterrestrial life. And, it is less than 600 million km (365 million mi) from Earth (at their closest distances).

SATURN AND ITS MOONS

The next planet from the Sun is Saturn, which has 62 known moons, and several are of great interest to astrobiologists. Like Jupiter, it is a gas giant, with an iron-nickel-rock-ice core, surrounded by liquid elemental hydrogen (H), liquid hydrogen (H_2), liquid helium, and then an atmosphere of hydrogen and helium, with small amounts of acetylene, ammonia, ethane, methane, phosphine, propane, and a number of larger hydrocarbon molecules. Saturn has a magnetosphere, but it is less intense than Jupiter's magnetosphere, and is also weaker than that of the Earth. So, it provides less protection to the atmospheres of its moons from the solar wind than does Jupiter. Although several planets have rings, Saturn's rings are the most extensive, and consist mainly of ice. Many of Saturn's moons also have considerable amounts of ice.

Mimas consists almost entirely of solid ice. While this might have microbes that have been preserved for billions of years, many of which may have originated elsewhere, there are probably no living organisms currently on this moon, with temperatures close to -160°C (-300°F). Enceladus initially looked like just another icy moon. But, imagery from a close flyby of the Cassini spacecraft in 2009 showed that there were more than 100 huge geysers on the surface of Enceladus that were spewing water vapor, hydrogen, salt, and ice crystals out into space. The composition is similar to that of comets. Some of the water vapor falls back to the surface as snow. As with some of the moons of Jupiter, there are indications of water oceans below the surface ice. The ice covering the ocean has been estimated to be only 30–40 km (18–25 mi) thick. If life exists in the ice or in the oceans, the geysers could be delivering some of those organisms to the surface (as well as out into space). Enceladus is certainly one moon that will remain a contender for a space mission to seek signs of extraterrestrial life. However, it is approximately 1.2 billion km (750 million mi) from Earth, and because its surface is so white and reflective, it is colder than most of the other moons, hovering around -200°C (-325°F). Tethys is irregular in shape and is similar to Mimas, in that it is almost entirely composed of water ice. It is nearly as reflective as Enceladus, and is almost as cold. As with Mimas, its ice might contain preserved organisms, but it is likely to be devoid of currently living organisms. Rhea is rocky and icy, and has a thin atmosphere consisting mainly of O_2 and CO_2. Therefore, it might have some interesting chemistry, but there are no indications of liquid water on the surface or in the interior.

Titan is the largest moon of Saturn, and the second largest moon in the Solar System. It is 50% larger than Earth's Moon. It is also the only moon that has a thick atmosphere, consisting of mainly nitrogen gas (N_2), with traces of methane, ethane, and nitrogen-rich organic compounds. Because it is shrouded in clouds, nothing was known about its surface, until the Cassini-Huygens spacecraft landed on the surface in 2005. Liquid lakes of ethane and methane were found, with temperatures around -180°C (-290°F). Life, as we know it, could not exist here, but different forms of replicative life forms might be possible. However, if there are alternative life forms here, it is unknown how they would be identified. So, although Titan is an interesting moon for many reasons, it would be difficult to plan for a mission to search for life there.

Hyperion is a potato-shaped moon that has a strange rotation. It has been described as a rubble pile or a sponge-like moon. It consists of mainly ice, but mixed with rocks and dust, that includes hydrocarbon compounds. Again, with this moon, preservation of organisms within the surface ice would be possible. Iapetus is composed of ice and rock, but with a very bright side and a very dark side. As with most of the moons in the Solar System, they have entrained one side towards the planets they orbit, and the other side away from their planets. Therefore, as they orbit, they have a leading side that collects space dust, and a trailing side that accretes much less dust. This is why the leading side is generally darker than the trailing side, especially when the surface is composed of mainly ice. Dione is composed of about 2/3 ice and 1/3 rock, and in many ways is similar to Rhea, although it has a series of icy cliffs on its trailing side. It has a shell of approximately 100 km (62 mi) of ice that is covering a 65 km (40 mi) deep liquid water ocean. The icy cliffs might indicate a thinner region through which the ocean might be accessed for study, although the thickness of the ice still presents currently insurmountable technical difficulties.

Phoebe is different from the other moons of Saturn in that it is thought to have come from the Kuiper belt, similar to Ceres. Although it is composed of ice and rock, it also has a large amount of carbon compounds. Both Ceres and Phoebe would be interesting to study to determine which organic compounds are present, and how these might relate to the origin of life on Earth and in the Solar System. However, Ceres is only 415 million km (250 million mi) from the Earth, while Phoebe is 1.2 billion km (750 million mi) from the Earth.

URANUS AND ITS MOONS

The seventh planet from the Sun, Uranus, is the third largest planet in the Solar System, and one of the two "ice giants," in addition to Neptune (Jupiter and Saturn are called "gas giants"). They are called ice giants because they have higher proportions of water, ammonia, methane, and hydrocarbon ices than the gas giants, although their atmospheres consist primarily of hydrogen and helium, similar to the gas giants. It has an extremely cold atmosphere with temperatures as low as -224°C (-370°F). It has a magnetosphere, as well as 27 known moons, of which six are relatively large. Puck and Miranda are comprised mainly of ice, although Puck is dark because of a dust layer. Ariel, Umbrel, Titania, and Operon are each about half rock and half ice, mixed with some ammonia and carbon dioxide ice. All of them are partially dark, indicating dust layers, with Ariel being the brightest of these moons. It is unknown whether any of them has any liquid water under the ice. Nonetheless, the ice mixed with dust layers may be good places to search for organic molecules and preserved organisms.

NEPTUNE AND ITS MOONS

Neptune is similar to Uranus in composition and size, although it is the fourth largest planet in the Solar System. Its atmosphere is primarily composed of hydrogen and helium, with other components mixed in, again similar to those found in the atmosphere of Uranus, but in different proportions. As with the other giant planets, Neptune has a magnetosphere, which protects its atmosphere, as well as those of the moons, from the solar wind. There are 14 known moons, including three that are larger than the others. Triton is the largest, with Proteus and Nereid being smaller. All three have an atmosphere composed primarily of nitrogen with small amounts of methane, carbon monoxide, carbon dioxide, and water ices. Temperatures are as low as -235°C (-390 °F). Because of the extreme distances to the moons of Neptune and Uranus, as well as their extremely low temperatures and somewhat complex atmospheres, sending spacecraft to these moons to seek signs of life would be technically challenging, very expensive, and would entail long time periods of operation because of the extreme distances.

PLUTO AND ITS MOONS

Pluto has been alternatively classified as a planet and a dwarf planet. Currently, it is designated as a dwarf planet, as are Ceres, Phoebe, and a few other known bodies beyond the orbit of Pluto. It originated in the Kuiper belt, and thus is different than all of the other main planets, and has more in common with Ceres in the asteroid belt. It has a surface that is primarily nitrogen ice, with some methane, carbon monoxide, and water ices. Additionally, it has water ice mountains, and a thick mantle of water ice. The water emerges from the interior as what appear to be shield volcanoes, that resemble the shield volcanoes of Hawaii. The volcanoes are extrusions of liquid materials, such as lava on Earth. But, on Pluto, the extruded liquid is water, which freezes on the surface, which is close to -185°C (-300°F). It also has pillow-like structures that are thought to be bubbling areas of liquid and frozen nitrogen. As with Ceres, and other Kuiper belt bodies, there are thiolins, a mixture of more complex organic compounds, on the surface. These compounds may have been important to the origin of life in the Solar System. Pluto has five known satellites, although Charon is by far the largest. The north pole of Charon also contains thiolins. It is composed of 55% rock and 45% ice, whereas Pluto is 70% rock and 30% ice. Both are candidates for the search for life, as well as the conditions for the origin of life in the Solar System, although their extreme distances from Earth, 7.5 billion km (4.7 billion mi), at their closest approach to Earth, make their study extremely challenging and prohibitively expensive.

A QUESTION OF LIFE

Life is diverse, complex, intricate, and ever-changing. Scientists have discovered and described only about 10–13% of the species thought to be living currently on Earth. Therefore, the vast majority of life on Earth is unknown. Most of that life is microscopic, and much of it is virtually inaccessible. We know that life exists in almost all environments on Earth from the bottoms of oceans to the tops of mountains, in boiling pools, encased in ice, and suspended in clouds. Additionally, life exists miles into the Earth's crust. In fact, microbes have been found in almost every place that scientists have looked. However, the organisms from these extreme environments are difficult to isolate and study. Thus, in many ways, it is difficult to catalog the majority of microorganisms and the diversity of life. Given the fact that over 99% of all species that have ever existed on Earth are now extinct, one would think that life is tenuous, ephemeral, and sensitive. However, some viruses, bacteria, and fungi survive under extreme conditions (of heat, cold, dryness, pressure, radioactivity, etc.). Therefore, some living things are extremely hardy. Some organisms can exist in states of suspended animation for years, decades, centuries, millennia, and longer. Thus, our definition of life, and the environments that support life, must be extended past what was thought only a short time ago. It appears that we have just scratched the surface of what comprises life, and where life might be found. In the mid-twentieth century it was thought that life could exist only within various narrow Earth environments, and that the only possible other location for life was in the warmer locations on Mars. Currently, it is widely accepted that life may exist on many of the planets, moons, and other bodies in the Solar System and beyond. The trick now is to find the best ways to detect life in these extreme environments. Astrobiologists are focusing their efforts on areas that contain water and ice. For the past several decades, there also have been many efforts to find signs of extraterrestrial life. The first requirement seems to be the presence of water, and this includes liquid and frozen water. No organisms have ever been found that lack a requirement for water. So, astrobiologists have first been searching for water on other bodies within the Solar System, and beyond. Water has been detected on our Moon, Mars, Mercury, asteroids, meteors, comets, and almost every dwarf planet and moon that has been examined. Several moons of other planets have thick sheets of ice, as well as deep liquid water oceans. Several meteorites that have struck the Earth contained several amino acids, as well as water. It was once thought that water and ice were rare in the Solar System. However, now we know that a huge amount of water and ice exist in many locations in the Solar System. Now, many studies are attempting to answer the question of whether life originated on Earth, or elsewhere and arrived on Earth on an extraterrestrial body. We and our colleagues study microbes in ancient ice and permafrost. These frozen substrates are unique biological repositories on Earth. They preserve living organisms and biological molecules by retarding degradative processes. Therefore, the state of preservation of organisms and molecules often is excellent even after centuries or millennia (or longer) of entrapment in the ice. During the past two decades, researchers have developed methods to study and catalog the organisms and biological molecules preserved in ice. While all of the studies to date have dealt with life on Earth, some of the methods are applicable to studies seeking life on other planets. They can be thoroughly tested on Earth before they are rocketed on their way to distant planets and moons in search of life there.

SOURCES AND ADDITIONAL READINGS

Price, P. B. 2007. Microbial life in glacial ice and implications for a cold origin of life, *FEMS Microbiology Ecology* 59:217.

Price, P. B. 2010. Microbial life in Martian ice: a biotic origin of methane on Mars. *Planetary and Space Science* 58:1199–1206.

Rogers, S.O, W.T. Starmer, and J.D. Castello. 2004. Recycling of pathogenic microbes through survival in ice. *Medical Hypotheses* 63: 773–777.

C. Tung, C., N. E. Bramall, and P. B. Price. 2005. Microbial origin of excess methane in glacial ice and implications for life on Mars. *Proceedings of the National Academy of Sciences of the United States of America* 102:18292 (2005).

LINKS

https://www.nasa.gov/jpl/the-solar-system-and-beyond-is-awash-in-water
https://en.wikipedia.org/wiki/Astrobiology
https://en.wikipedia.org/wiki/List_of_missions_to_Mars
https://en.wikipedia.org/wiki/Mercury_(planet)
https://en.wikipedia.org/wiki/Venus
https://en.wikipedia.org/wiki/List_of_missions_to_Venus
https://en.wikipedia.org/wiki/Moon
https://en.wikipedia.org/wiki/Mars
https://en.wikipedia.org/wiki/Ceres_(dwarf_planet)
https://en.wikipedia.org/wiki/Gas_giant
https://en.wikipedia.org/wiki/Moons_of_Jupiter
https://en.wikipedia.org/wiki/Moons_of_Saturn
https://en.wikipedia.org/wiki/Ice_giant
https://en.wikipedia.org/wiki/Moons_of_Uranus
https://en.wikipedia.org/wiki/Moons_of_Neptune
https://en.wikipedia.org/wiki/Pluto
https://en.wikipedia.org/wiki/Kuiper_belt
https://www.popsci.com/60-tons-cosmic-dust-fall-earth-every-day/
https://www.space.com/35469-solar-system-habitable-icy-worlds-infographic.html

19 Disappearing Ice— Global Climate Change

"We are accidents waiting to happen"

Thom Yorke

POLLUTION AND MELTING

During the Industrial Revolution, which began in the mid-18th century and accelerated in the 19th century, many innovative forms of manufacturing, building, agriculture, and transportation were invented and put into practice, resulting in economic and societal changes, as well as many environmental changes. The use of fossil fuels shot up enormously, and by the late 19th and early 20th centuries the environmental changes were becoming toxic and pervasive (Fig. 19.1). Polluted air, waterways, lakes, oceans, and land areas were becoming commonplace, and there were significant health hazards to large populations of people, as well as to plants, animals, and other organisms. While some controls were enacted to curb some obvious waste problems, other emissions went unabated, and some were unrecognized. Among those that were unrecognized or ignored were greenhouse gas emissions. However, as early as the late 19th century, scientists discovered that some of the gases emitted by industry could cause warming of the atmosphere. During that same period, others were chronicling the retreat of glaciers in the Alps and elsewhere. It would take decades to connect the two phenomena. But by the 1960's, climate scientists reported that the overall temperature on Earth had been rising during the first half of the 20th century, and connected it with a rise in greenhouse gases, primarily carbon dioxide and methane (Fig. 19.1). While some proposed that the rise was from volcanic eruptions and other natural sources, many linked it with human activities. Even scientists and officials of fossil fuel (e.g., gas, oil, and coal) industries realized in the 19th and 20th centuries that the production and utilization of their products released large quantities of carbon dioxide and methane into the atmosphere, and these could lead to warming of the atmosphere. However, they fought against any measures that would cause them to cut production or install equipment to reduce the emissions, primarily for economic reasons.

Some scientific studies have measured the increases in these gases and calculated the amount of temperature rise, predicting dangerous global temperature increases, more severe storms, increases and decreases in precipitation, glacier recession, sea level rises, and degradation of many habitats worldwide. These

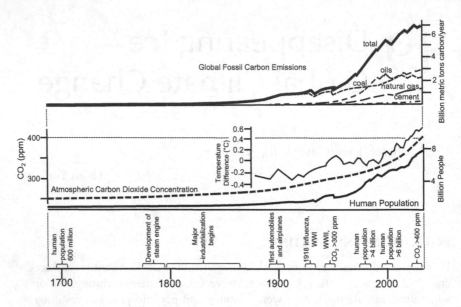

FIGURE 19.1 Factors affecting global climate change from 1700 CE to the present. The upper graph shows the amounts of carbon emissions from various fossil fuels. The middle graph relates the changes in atmospheric carbon dioxide, global temperatures, and human population. The lower bar indicates some of the pertinent milestones during this time period.

predictions have already come true. As carbon dioxide emissions continued to rise, as global temperature records were set, as sea levels rose inundating island and coastal communities, as permafrost melted, as glaciers continued to recede, as coral worldwide died, the debates about global climate change continued. In the 1980's and early 1990's, some models seemed to indicate that the amount of temperature rise was lower than predicted by the models. However, the measurements were primarily from temperatures on land. Then, in the mid-1990's, more measurements were collected from the oceans, and the numbers agreed closely with the models of energy increases from carbon emissions. The oceans were heating up more rapidly than the land. The water on Earth was acting like a huge heat sink. It was discovered that the temperature of the oceans has risen by an average of 0.13°C (0.23°F) per decade during the past century, a total rise of 1.5°C (2.7°F), and recently the temperature has been increasing at a faster rate. When you think of the enormous size of the ocean, it is almost unimaginable how much energy it has taken to increase the temperature of all of that water, an estimated 1.335 billion cubic kilometers (320 million cubic miles). This has several implications. First, more heat in the oceans results in more evaporation and more clouds, which leads to shading some areas of the oceans and land, but also leads to more precipitation, including rain and snow. Second, more energy in the oceans and atmosphere causes more intense and larger tropical storms, hurricanes, and cyclones, and potentially can change ocean currents. Third, it leads to more melting of sea and coastal ice, which leads to rises in worldwide sea levels. Fourth, when liquids are heated, they expand, which also leads to increases in sea levels, from thermal expansion. The increases in precipitation and sea level have already been documented. Fifth, increases in ocean temperatures causes stress on sea life, such that some species have moved further north, and other species have declined in numbers or have gone extinct. Corals have died worldwide, and fish are moving north in the Northern Hemisphere. Sixth, the melting of the ice, and the warmer water temperatures have encouraged the growth of many types of microbes, some of which are pathogens.

RECEDING GLACIERS AND ICE SHELVES

While records have been kept on glaciers in Europe for centuries, detailed examinations of glaciers and ice fields in Antarctica have been undertaken only during the last several decades. Regardless of when the measurements were made, the conclusions have been the same: Glaciers, ice fields, sea ice, and ice shelves have been receding, and some have simply disappeared (Fig. 19.2). Several glaciers in Europe have received the greatest scrutiny. The Rhône glacier in the Swiss Alps is the primary source of water for the Rhône River, which flows into Lake

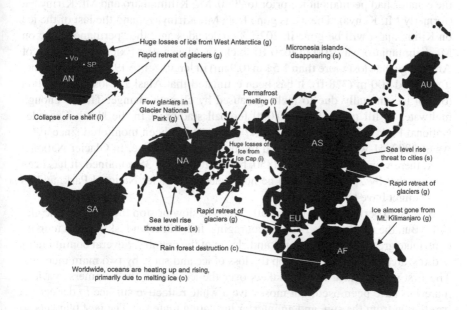

FIGURE 19.2 Map of the world showing some of the major changes causing or resulting from global climate change. Letters are: (c) destruction of vegetation that sequesters carbon; (g) melting and retreating glaciers; (i) melting ice, ice shelves, and permafrost; (o) warming oceans, melting ice, causing species declines; (s) sea level rise, partly due to melting and partly due to thermal expansion, threatening cities and islands. Continents are: AF—Africa; AN—Antarctica; AS—Asia; AU—Australia; EU—Europe; NA—North America; SA—South America.

Geneva, on the border between France and Switzerland. During the 1600's, the glacier was fairly stable, and its terminus was near the town of Gletsch (gletscher is the German word for glacier). During the late 1700's and early 1800's (corresponding to the beginning of the Industrial Revolution in Europe), it began to recede. By 1856, it had receded by about 300 m (985 ft). By 1880, the terminus was more than a kilometer back from where it had been in the 1600's. By 1945, the terminus was 2 kilometers (1.2 mi) away from the town of Gletsch, and 200 m (656 ft) in elevation above its position in 1602. By 2014, it was more than 3 km (1.86 mi) from Gletsch, and 700 m above the town. At its greatest extent, it was nearly a kilometer wide and 300 m (985 ft) deep. Today, it is much smaller. The amount of meltwater and the number of microbes released from 1800 to 2000 was enormous. And, all of that water flowed into Lake Geneva, which is the source of drinking water for more than one million people.

Other glaciers in the Alps also have been examined and measured. The Aletsch glacier, also in Switzerland, currently is 23 km (14.3 mi) in length, covering almost 82 km^2 (31.6 mi^2), and consists of more than 15 km^3 (3.6 mi^3) of ice. Since 1870, it has retreated more than 3 km (1.9 mi), and lost more than 300 m (984 ft) in height. Similarly, the Morteratch glacier has retreated 2.5 km (1.6 mi) since 1860. This trend in glacier recession is worldwide. In Africa, two mountains near the equator had permanent ice prior to 2000, Mt. Kilimanjaro and Mt. Kirinyaga (formerly Mt. Kenya). The ice is gone from Mt. Kirinyaga, and the last of the ice on Kilimanjaro will be gone in 2020. Two decades ago, the "permanent" ice on Mt. Kilimanjaro was 200 m (650 ft) thick. In Alaska, the Exit glacier, south of Anchorage, has lost more than 1.5 km (0.9 mi) in length since 1950. In 2016 alone, it retreated 100 m (328 ft). It has been estimated that Alaska is losing 75 billion tons of ice annually due to melting caused by climate change. This is enough meltwater to fill more than 100,000 football stadiums! In the North Cascades National Park in Washington State, 47 glaciers have been monitored since 1967. As of 2007, 41 were retreating and the other six were gone. In Glacier National Park, there were 150 glaciers in 1850, but by 2018, only 35 remained. It has been forecast that by 2030, there will be no glaciers in Glacier National Park, and its name might have to be changed.

Several experiments have been attempted to slow or stop the loss of these glaciers. But, the results have been discouraging. In Switzerland, skiing and tourism contribute greatly to the economy and charm of the country. Several communities and researchers have tried to stop the loss of ice and snow by two main methods. The first was to place large blankets over the ice during the summers. Various materials have been used, but most have a white reflective surface to deflect the irradiation from the sun, and an interior insulating material. The best blankets so far tested have slowed summer melting by as much as 80%, but none have stopped it completely, so the glaciers continue to retreat. Also, this only works for relatively small areas. You can imagine trying to cover the entire length of the 23 km (14 mi) long Aletsch glacier with blankets. The other main method that has been attempted is to cover the glaciers with artificial snow. As with the blankets, this can slow melting and retreat of the glaciers, but it has not halted the retreat of the glaciers. And, as with blankets, it can only be used in small areas that are accessible to snow-making equipment.

The ice mentioned above is only a very small proportion of the ice on Earth. Approximately 90% of the fresh water on Earth is frozen in ice fields and glaciers. About 8% of this ice is in Greenland, and 91% is in Antarctica (see Table 4.1). They, too, have experienced huge losses of ice during the past century, and the melting is accelerating. Currently, Antarctica is losing 278 billion tons of ice annually, up from losses of 44 billion tons annually in 1980. This is equivalent to nearly 550 trillion pounds of ice! Studies by several different groups of microbiologists have determined that there are usually from 10 to 10,000 (or sometimes more) cells per ml in this ice, which represents a release of 10^{18} to 10^{22} cells (many of them viable) annually from the melting of ice in Antarctica. If the other ice worldwide (including from Greenland) is added to this, then the numbers could be about 10% higher. On a daily basis, this means that on average, approximately 10^{15} to 10^{19} microbes are being released from melting ice. These releases are increasing due to warming temperatures, especially near the poles.

While most of the ice loss has been from melting glaciers and ice shelves in West Antarctica, increasing losses from East Antarctica have scientists worried, because it was thought to be resistant to melting. West Antarctica is much lower than the highlands of East Antarctica. Thus, regions of West Antarctica are melting much faster than most areas of East Antarctica. Some of the glaciers are speeding up in their movement towards the sea, and the ice sheet shows signs of instability. Currently, one of the most concerning areas is the Thwaites glacier in West Antarctica. It begins far inland as an outlet of the West Antarctic Ice Sheet. In recent years, it has been showing signs of increased movement. It is held back primarily by an ice shelf, and the rough surfaces below the glacier. However, indications are that these could be overcome by further warming of the ocean and by surface warming of the glacier. If the Thwaites glacier collapses, sea levels could rise by more than 3 m (10 ft) in a few decades. Imagine all of the coastal cities and islands that would be affected by this level of rise in the oceans around the world. It, alone, has lost 600 billion tons of ice since 1980. Parts of this region collapsed into the ocean over 100,000 years ago, which raised the oceans dozens of meters. Even if the portion closest to the ocean collapsed today, the 3 m (10 ft) rise in sea level would cause damage to coastal cities around the world, especially during high tides and storms. It currently is in danger of collapsing. Many regions, especially islands, would become uninhabitable. In 2018, a huge underwater cavern was discovered under the glacier. It covers an area the size of Manhattan, and is over 300 m (1,000 ft) tall. The cavern was not there three years before! This has startled scientists, leading them to believe that the end of the glacier could soon collapse. If it does, this could trigger increased movement and further collapses, leading to rises in sea level worldwide. Glaciers are receding rapidly in Greenland, also. Huge pieces of the Helheim glacier have been collapsing. The Jacobshaven glacier drains about 6.5% of the ice from Greenland, and its movement and recession have increased during the past several decades.

Sea ice in the Arctic is rapidly declining (Fig. 19.2). In the 1990s, sea ice in the Arctic covered 15 million km^2 (5.8 million mi^2) in the Winter. By 2018, the maximum extent was down to 12 million km^2 (4.6 million mi^2). During the same time period, the sea ice during the Summer has decreased from 7 million km^2 (2.7 million mi^2) to 5 million km^2 (1.9 million mi^2). Ice shelves are present in the Arctic and Antarctic. The ice originates on land, and flows out into the ocean, where it floats in the cold saltwater. This differs from sea ice, which is frozen saltwater. The ice shelves consist mainly of fresh water, although some saltwater from the ocean freezes to its lower surfaces of the ice shelves. Part of the loss of ice from Antarctica has been from ice shelves, and in many cases, these slow the movements and collapses of glaciers. Slabs of ice the size of small states (e.g., Rhode Island and Delaware) have broken free from the ice shelves, and more of these events are expected to occur in the near future. Originally, the Larsen B ice shelf covered an area of 85,000 km^2 (33,000 mi^2). Between 1995 and 2017, several large pieces broke free from the shelf, and by 2018, it covered an area of 67,000 km^2 (26,000 mi^2). The Wilkins Ice Shelf is located between Charcot Island, Latady Island, and Alexander Island, offshore from the Antarctic Peninsula. In the early 1990's, it covered an area of 16,000 km^2. However, pieces began breaking off by the late 1990's, and by 2009, it had lost more than 3,000 km^2 (1,160 mi^2). Glaciers on the Antarctic Peninsula, which is the piece of the continent that extends the furthest northward, also are losing much of their mass. Some glaciers have receded several kilometers in just a few decades. Some of the research stations and cabins that were at the bases of glaciers, now are many kilometers from the glacial termini, due to rapid glacial recession.

GREENHOUSE EFFECT

The major cause of melting ice worldwide is the increase in atmospheric, land, and ocean temperatures, which has been greatly accelerated by human activities, due to the enormous releases of greenhouse gases into the atmosphere. The greenhouse effect begins when light passes through the atmospheric gases, such as carbon dioxide and methane. The photons then strike land and water, and transfer their energy into molecules on the land and water, where their energy is mostly converted to heat in the form of infrared radiation. This is similar to when photons enter a greenhouse or your car. The photons go through the glass, but their energy is converted into heat when they strike the surfaces in the greenhouse or the interior of your car. However, the heat is stopped by the glass. In the atmosphere, the greenhouse gases do not allow the infrared heat energy to radiate back into space, and thus the atmosphere and land heat up. It is similar for areas with water, where the photons also are converted to heat. However, at certain angles, some of the photons are reflected off the surface of the water. For ice, and any white surface, a majority of the photons are reflected back into the atmosphere, and many are reflected back into space. This phenomenon is called albedo. Snow and ice-covered surfaces on the Earth have the highest albedos, and thus reflect large proportions of the sun's rays back into the atmosphere and out into space.

A dark part of the land has a low albedo. As the ice melts, the total albedo of the Earth decreases, and more of the rays of the sun are absorbed, converted to heat, and thus warm the land and the water. Eventually, this causes a continuing feedback loop, which leads to more melting and more heating. Melting of the glaciers on land also leads to increases of sea level, with the accompanying loss of land area. Scientists already have noted that a large proportion of the heating on Earth is occurring in the oceans. Therefore, while the ice reflects the light and heating, the additional water absorbs that light and converts it to heat. This is yet another part of the loop that increases the rate of heating on Earth. The complexities of the current situation of the melting ice are of great concern because it is difficult to know how and where to attempt to slow down the melting. Covering the ice with a blanket will not work. It would be like trying to stop a moving oil tanker by exhaling as it approaches. An obvious beginning would be to stop releasing greenhouse gases into the environment by curtailing the heavy reliance on fossil fuels.

MISCONCEPTIONS

There are many misconceptions about climate change, including the fact that daily or weekly weather is only one component of climate, which is the long-term characteristic of the weather patterns in an area. Also, there is confusion about ice in the water versus ice on land, and the effects of melting for both. When sea ice melts, it does not change the depth of the water. This is similar to when an ice cube melts in a glass of water. After all of the ice has melted, the level of the water is the same as when the ice cube was floating in the water. However, fill a glass of water to the brim, and then add ice. The liquid in the glass overflows and the ice melts. Or, fill one glass with water, and melt an ice cube in another glass. Once the ice cube has melted, pour the water into the other glass that was filled with water. The level of the water definitely changes. So, if ice on land melts, and the meltwater ends up in the ocean, the level of the ocean will rise, especially when large amounts of ice melts into the ocean. For example, if all of the land ice on Earth melted, it has been estimated that sea level would rise more than 70 m (230 ft) worldwide.

Another misconception is that all of the plants will grow better because of the higher concentrations of carbon dioxide in the atmosphere. Therefore, adding carbon dioxide to the atmosphere will make plants grow better. It is not this simple. While it is true that the concentrations of greenhouse gases in the atmosphere of Earth have been higher in the past, and in general, with all other factors remaining constant, plants can grow faster when they receive more carbon dioxide. However, when carbon dioxide levels in the atmosphere increase, temperatures increase, and all plants do not like hotter temperatures. The fact is that humans, and their agriculturally manipulated plants and animals, were not present during times in the past when carbon dioxide levels and temperatures were warmer. That is, the amount of carbon dioxide and methane in the atmosphere have never been this high during the past 200,000 years, which is the entire length of time that humans have existed (Fig. 19.3). The agricultural animals and plants also were

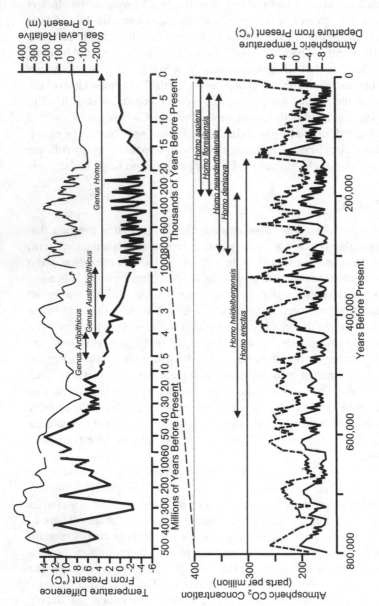

FIGURE 19.3 World temperatures and sea levels during the past 500 million years, and carbon dioxide levels during the past 800,000 years, overlaid with the duration of existence for selected hominin species. The top graph indicates the extent of the hominin species that have existed for the past 5 million years. During that time, sea level increases and decreases have sometimes been large (plus and minus 100 m, 328 ft), while temperatures have decreased by as much as 8°C (14°F), or increased by as much as 2°C (3.6°F). Prior to that time, sea levels rose much more dramatically (up to 400 m, 1,300 ft) during some periods, and temperatures were much higher (up to +15°C, 27°F) during some periods.

not present at that time. Humans have bred these plants and animals to grow in certain places on Earth, and all of them were bred by humans only during the past 11,000 years. When climate patterns change, rainfall is changed, and winds increase, which affect the plants, and as the temperatures increase, many plants die, or fail to reproduce because they cannot produce seeds or pollen, which are temperature-sensitive processes. Rapid changes kill many of the plants because they cannot move. Also, many plants need a cold period in order to germinate, produce flowers, reproduce, etc., in a process called vernalization. If the cold periods are reduced, the plants and their seeds are not properly vernalized, and many of them will fail to grow properly, or simply die. Animals die if they are too hot or do not have enough water or food. Climate change already has begun to alter all of these variables.

Plants cannot move (except as seeds or pollen), but animals and microbes can move, either by walking, flying, swimming, or by the action of wind. During the past few decades, fish in the oceans have begun to move further north in the Northern Hemisphere and south in the Southern Hemisphere to find suitable water temperatures. However, what they eat may not move at the same rate or in the same direction. Corals have been dying worldwide, possibly due to the warmer water temperatures. Trees in the mountains have begun to die at lower elevations, and seedlings are being found higher in those mountains. However, eventually, their distributions can only increase in elevation until they reach the tops of the mountains. Another phenomenon that has been occurring is that the distribution of pathogens, as well as some of the animals that carry them have been changing. Mosquitos and other arthropods, have been moving northward in the Northern Hemisphere, such that malaria, dengue fever, zika, chikungunya, and other diseases are now popping up far north of their historical ranges. This includes not only human diseases, but diseases in other animals and plants. Outbreaks, epidemics, and pandemics of various diseases have been increasing.

GLOBAL CLIMATE AND HOMININS

Our species, *Homo sapiens*, emerged approximately 200,000 years ago from another species, *H. heidelbergensis* (Figs. 19.3 and 19.4). This species also led to the establishment of other species, including *H. neanderthalensis* and *Homo floresiensis*. Furthermore, *H. heidelbergensis* had originated from branches of *H. ergaster* and/or *H. erectus*, which had diverged from a branch of either *H. habilis* or *H. rudolfensis*, both of which diverged from a branch of a member of the related genus *Australopithicus* (probably from the species *A. africanus*, represented by the famous "Lucy" fossil), approximately 2.5 million years ago. These had diverged from a branch of another genus, *Ardipithicus*, approximately 4.3 million years ago. These were near the base of the branch that separated the hominins (human-like species) from the great apes (e.g., apes, chimpanzees, and bonobos; part of the hominid group). During the entire time that humans

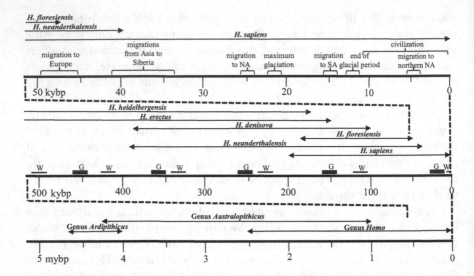

FIGURE 19.4 Timelines outlining some of the major events in human evolution compared to migration and glaciation events. Glaciation events influenced human migrations. Hominin speciation and extinction events may also have been influenced by climate changes. Evidence for some of these migrations may still exist in the Greenland ice, Antarctica ice, permafrost (and elsewhere). Bars indicate the glacial (G) and warmest (W) periods (within interglacial periods) during the last 500,000 years.

(*H. sapiens*) have walked the Earth, and when all other *Homo* species were present at the same time, the global temperatures have decreased by as much as 4–8°C (7–14°F, during ice ages) and risen by as much as 2–3°C (3–5°F, during warm periods), compared to current temperatures. These have been accompanied by sea level changes, with lows of approximately -100 m (-328 ft) to highs of +100 m, compared to current levels. The temperature diversions and sea level changes led to significant disruptions and migrations of human populations. Relatively minor fluctuations in temperature occur regularly in the ice core record (Fig. 19.3), such as the Medieval warming period (from about 950 to 1250 AD) in the Northern Hemisphere, which was 1–2°C (2–3°F) warmer than today, and the "Little Ice Age" (from 1300 to 1850 AD), which was from 1–2°C cooler than present temperatures. Although these were relatively minor fluctuations, they caused humans to make adjustments in their activities, and they were somewhat upsetting for those people, including increases in crop failures, famines, and disease epidemics.

Major temperature fluctuations, such as warming periods with much higher temperatures, and true Ice Ages, including global freezing periods, with much colder global temperatures have occurred on longer timescales. These caused major problems for humans, and some populations died out completely during these times. For example, the higher temperatures and sea levels that were experienced by *Ardipithicus* and *Australopithicus* individuals have never been experienced by an individual of any *Homo* species, including any member of *H. sapiens*.

Mean world temperatures were up to 4°C (7°F) or more above current levels, and sea levels were sometimes more than 300 m (984 ft) higher than today. During the past 800,000 years, carbon dioxide levels were never higher than 300 ppm (parts per million) in the atmosphere. In 2018, carbon dioxide levels surpassed 400 ppm, and continue to rise. Methane levels are on the increase, and have never been higher during the same time period. During the past 800,000 years, methane concentrations in the atmosphere never exceeded 800 ppb (parts per billion). As of 2019, the concentration in the atmosphere is above 1,860 ppb, and recent more rapid increases have exceeded predictions. While this greenhouse gas is in lower concentrations than carbon dioxide, it is a much stronger greenhouse gas than is carbon dioxide, so this has scientists around the world very worried. It also is converted to carbon dioxide in the atmosphere, and therefore contributes on long timescales to the total levels of greenhouse gases. Drilling and fracking for fossil fuels, processing fossil fuels, transporting fossil fuels, leaky valves, agricultural practices, warming oceans, melting ice, and melting permafrost all contribute to the increases in atmospheric methane. Land and ocean temperatures also continue to increase, which will continue to accelerate these releases. The Earth is warming. The ice is melting. Sea level rise is affecting major cities and many islands. Stopping it might be impossible. Slowing it is possible. The problem is clear and right now there are only a few actions that will slow this process.

OTHER PLANETS AND THE FUTURE OF EARTH

The study of other planets may help to determine the future of Earth. Our planet is in the so-called "Goldilocks Zone," which is not too hot, not too cold to support life. Its atmosphere moderates the temperature at the surface so that life is possible. It also has a lot of liquid water, which is required for life, at least life as we know it. During the past 50 years, we have learned that most planets and moons also have water, and some moons are almost all ice! Also, we have learned that planets with magnetic poles produce a magnetic field around them (called a magnetosphere) that deflects the energetic charged particles coming from the Sun, called the solar wind. This deflection allows the planet to retain an atmosphere. Earth's atmosphere, which is protected by a strong magnetosphere, contains mainly nitrogen, which is used by nitrogen-fixing bacteria to add nitrogen to organic molecules. It also contains a large amount of oxygen, which is needed by aerobic organisms, such as humans, animals, plants, and many types of microbes. It also contains carbon dioxide, which is used by some photosynthetic organisms to make organic molecules, and also includes several other minor gases (e.g., methane). This was not always the case. On ancient Earth, carbon dioxide was in high concentrations, and oxygen levels were near zero. Atmospheric temperatures also were much higher than today, with some estimates of 70°C (160°F) at the surface. At that time (4.0 billion years ago), there already were oceans, but probably no permanent ice, and life had already gotten a start. Carbon dioxide levels decreased, and oxygen levels increased, about 2.7 billion years ago. The first major relatively permanent ice and permafrost appeared at that time. Since that time, in general, when carbon dioxide levels have risen, and oxygen levels have decreased, temperatures have increased. Conversely, when carbon dioxide levels have decreased, and oxygen levels have increased, temperatures have tended to decrease, sometimes causing ice to form from pole to pole, in so-called "Snowball Earth" events. After the last worldwide freezing event that occurred from about 600–700 million years ago, there was a major expansion of the number of different species on Earth, called the Cambrian Explosion. This coupling between the end of major ice events and evolution is interesting, although the actual process of the speeding up of evolutionary processes after these major ice periods remains unknown.

Consider the planets Mercury and Venus. Mercury has a magnetosphere, but it is only 1% of the strength of Earth's magnetosphere. This has caused it to maintain a thin atmosphere, consisting of hydrogen and helium, mainly coming from the Sun, and some oxygen. It may have a small amount of ice at its north pole, which constantly is pointed away from the sun. Next from the Sun is Venus, which has been described as a hellish world. Early Venus (4.5 to 4.0 billion years ago) probably was much like the Earth, including having oceans. But, it was much closer to the Sun (currently, 76 million miles versus 93 million miles for Earth). This caused much more evaporation of the oceans until they were completely gone. At the same time, the tug on Venus from the Sun caused more distortion of the surface, which caused more vulcanism. The volcanos spewed out enormous clouds of gases, which included carbon dioxide and sulfur compounds. The large amounts of carbon dioxide-caused extreme greenhouse effects, such that the surface temperature of Venus is now approximately 480°C (900°F), which is hot enough to melt some metals, some of which become gases, and precipitate as metallic snow. Some of the mountains on Venus look snow-covered, but they are actually lead covered, from lead "snow". The mass of the atmosphere is also 90 times that of Earth's atmosphere, so at the surface, almost everything on Earth would be crushed on Venus. The first four Russian spacecraft that landed on Venus were crushed before they landed. Several others landed successfully, but sent back video images for only 20 to 120 minutes before they were crushed and all signals from them stopped.

The Earth was lucky to have avoided the extreme greenhouse effects that are present on Venus. Hopefully, our current situation will not be a repeat of what is found on Venus currently. Several things have helped the Earth to avoid a runaway greenhouse effect. First, carbon dioxide has dissolved to a certain extent into the oceans. Second, many microbes and plants can take up the carbon dioxide and attach it to organic molecules that it uses for many purposes. Larger organisms can then use those carbon compounds for their survival. Much of the ancient carbon is in the form of so-called fossil fuels, being remnants of long-dead plants and other organisms. Some organisms use the carbon from the carbon dioxide bound to calcium to build protective calcium carbonate shells around their soft body parts, including corals and bivalves (e.g., clams and scallops). Red and green algae, as well as plants have chloroplasts (which were derived from cyanobacteria) that can attach (or fix) carbon from carbon dioxide onto larger organic molecules. Many protozoans also have this capability, although they gained this ability from the red and green algae. Therefore, from some biological processes, as well as from some non-biological processes, the amount of carbon dioxide in the atmosphere has been controlled for billions of years to a great extent by the actions of organisms. However, the Earth still is in a location in the Solar System where a greenhouse effect could heat the planet to the point where it would make things difficult or impossible for many types of life to survive (including humans). Out further from the Sun, this is less likely to occur, because the solar irradiation is less, and the temperatures are much lower, such that water and many other substances freeze, and atmospheres are generally composed of different gases, and, except for Titan and the gas and ice giants, the atmospheres are thin.

PLUSSES AND MINUSES

The fact that many types of microbes and even some multicellular organisms can survive being frozen in ice for hundreds to sometimes millions of years, is clear evidence that environmental ice acts as a repository of viable organisms, but also serves as a repository of genetic information on Earth as well as in other locations in the Solar System. While most organisms are harmless, some are pathogenic on plants, animals, and other organisms. With increases in ice melting on Earth, the potential for isolated disease outbreaks, as well as epidemics and pandemics also increases. On the other hand, some are likely to be beneficial to plants, animals, and other organisms. Therefore, there may be beneficial, or at least informative, organisms that are being lost as melting rates increase worldwide. For example, many plant viruses survive well through freeze and thaw cycles, and some may survive for hundreds of years in environmental ice. Some animal (including human) viruses also survive freeze and thaw cycles, including influenza A virus, which annually causes tens of thousands of human deaths in the US alone. If these and other viruses can survive in ice for long periods of time, they can emerge and infect a population that has no immunity to the particular strain of virus. Beneficial organisms may also survive freeze and thaw cycles in environmental ice. These organisms can be beneficial to people in a number of ways, including acting as part of their protective microbiome, being important in producing foods, aiding in agriculture, making valuable pharmacological products, controlling pathological or destructive organisms, and many others. These resources are disappearing rapidly, before they have even been investigated. The rapid rate of melting is eliminating multitudes of potentially beneficial microbes. But, on the minus side, the accelerated release of pathological organisms is also increasing, posing dangers to human populations, especially those that are near the glacial outlets, or those that utilize glacial meltwater for drinking, cooking, and agriculture.

Another benefit of preserving as much ice on Earth as possible is that it has a high albedo. The land, oceans, lakes, forests, and meadows absorb from 80–95% of the sunlight that strikes them, and most is converted into heat, and a small portion is used to fix carbon (photosynthesis in plants and microbes). On the other hand, snow and ice reflect from 30–85% (depending on the dust and soil content) of the light back into space. Between land ice and sea ice in Antarctica and the Arctic, a large proportion of the sunlight is reflected away from the Earth, and therefore, it does not contribute much to heating of the land, oceans, or atmosphere. As greenhouse gases continue to rise, largely due to human activities, more ice and permafrost melts. This releases more organisms, as well as more greenhouse gases. It also increases the rate at which the Earth absorbs solar radiation, and converts it to heat. This leads to a cycle of more melting and more heating. Other than trying to decrease the release of greenhouse gases by reducing the burning of fossil fuels, no current technology exists to stop, or even slow down, this process.

An immediate issue that must be addressed very soon derives from the fact that approximately 40% of the people in the world (3 billion people) depend on glacial meltwater for their drinking water and livelihoods. Around the Himalayas alone, more than 1.5 billion people depend on glacial meltwater. It has been estimated that by 2100, one-third of the glaciers in the Himalayas will be gone, thus threatening the lives of those people. As was mentioned in a previous chapter, many people in Europe depend on the water coming from the glaciers in the Alps. These glaciers also are retreating and disappearing. Europeans already are planning for life without so many glaciers. Throughout Europe, the Western US and Canada, glacial meltwater is used for drinking water, to irrigate crops, for industrial endeavors, for hydroelectric power, and many other purposes. As climate change has affected precipitation patterns, snowfall amounts, glacial melting, and temperature rises, natural disasters have resulted, including droughts, extreme rainfall events, crop failures, forest fires, home destruction, mud slides, etc. Some communities and countries are beginning to respond to the economic and personal losses. Indonesia is planning to move its capital city. Others continue to ignore the dangers. Study of ice in polar regions has helped to recognize the causes of these disasters, and still there is more to be learned, with the hope that a solution can be found through this research.

SOURCES AND ADDITIONAL READINGS

Augustin, L., C. Barbante, P.R. Barnes, et al. (55 authors). 2004. Eight glacial cycles from an Antarctic ice core. *Nature* 429: 623-628.

Castello, J.D., and S.O. Rogers. 2005. *Life in Ancient Ice*. Princeton, NJ: Princeton University Press.

Cheng, J., J. Abraham, Z. Hausfather, and K.E. Trenberth. 2019. How fast are the oceans warming? *Science* 363: 128-129.

Petit, J.-R, J. Jouzel, D. Raynaud, N.I. Barkov, J.-M. Barnola, I. Basile, M. Bender, J. Chappellaz, M. Davis, G. Delaygue, M. Delmotte, V.M. Kotlyakov, M. Legrand, V.Y. Lipenkov, C. Lorius, L. Pépin, C. Ritz, E. Saltzman, and M. Stievnard. 1999. Climate and atmospheric history of the past 420,000 years from the Vostok ice core, Antarctica. *Nature* 399: 429-436.

Reich, D., N. Patterson, M. Kircher, F. Delfin, M.R. Nandinemi, I Pugach, A.M.-S. Ko, et al. 2011. Denisovian admixture and the first modern human dispersals into Southeast Asia and Oceania. *Am. J. Hum. Genet.* 89: 516-528.

LINKS

https://climate.nasa.gov/
https://desdemonadespair.net/2015/06/graph-of-day-carbon-emissions-and-human.html
https://earthobservatory.nasa.gov/world-of-change/LarsenB
https://en.wikipedia.org/wiki/Global_warming
https://en.wikipedia.org/wiki/Northern_Ice_Field_(Mount_Kilimanjaro)
https://en.wikipedia.org/wiki/Paleothermometer
https://en.wikipedia.org/wiki/Retreat_of_glaciers_since_1850
https://en.wikipedia.org/wiki/Thwaites_Glacier
https://fox2now.com/2018/02/26/nasa-releases-time-lapse-of-the-disappearing-arctic-polar-
   ice-cap/
https://oceanservice.noaa.gov/facts/sealevel.html
http://www.antarcticglaciers.org/glaciers-and-climate/what-is-the-global-volume-of-land-
   ice-and-how-is-it-changing/
http://www.bbc.com/future/story/20190612-the-poisons-released-by-melting-arctic-ice
https://www.climateprediction.net/climate-science/glossary/greenhouse-effect/
https://www.dw.com/en/polar-ice-sheets-melting-faster-than-ever/a-16432199
https://www.handprint.com/LS/ANC/evol.html
https://mashable.com/article/himalayas-glaciers-climate-change/

20 Epilogue

"There is no planet B"

Barack Obama (among others)

IGNORING THE DANGERS

This Winter, as in past Winters, dozens of ice fishermen walked far out on the ice of Lake Erie. And, as in some years, they heard a crack and felt a jolt, and in a split second, they were on a large slab of ice that was no longer connected to the land. Some jumped into the widening crack and swam to the ice still connected to the shore. Luckily, none of them died (although some years, some do), but they were very wet and cold. The others waited for first responders to take them to safety. The first responders train for this every year, because it happens almost every year. Also, along the northern tier of states in the US, numerous people lose their cars, and some lose their lives, by driving or walking on thin ice to go ice fishing. Of course, all of them realize that it might be dangerous, but they ignore the danger. They think that it will not happen to them. We are in an analogous situation today. The evidence for human-caused acceleration of global climate change is firmly established. It is clear that it has led to sea level rise, and will cause more rises in sea level. It will lead to increases in human displacement, extreme weather events, diseases, wars, and deaths. Yet, the efforts to slow climate change have been feeble or nonexistent, so far, and some governments refuse to recognize that it is real. If it is allowed to continue, runaway global heating, ice melting, and sea level rise will be impossible to slow or stop. And in that case, no first responders will be able to help.

Glacial ice from Antarctica, Greenland, and elsewhere has already yielded a treasure trove of information about past climate, volcanic events, microbial preservation, microbial reintroduction, genome recycling, and others. Ice core data, and ground, ocean, and satellite data, have clearly proven that global climate change is a real phenomenon, and that it has occurred many times in the past, and will occur into the future. However, the ancient ice core data also shows that the current concentrations of greenhouse gases in the atmosphere, especially carbon dioxide and methane, are at unprecedented levels, which is linked to human activities. They are 50–100% higher than they have been at any time during the past 800,000 years. The synchrony between these gases and global temperatures has been measured, graphed, and confirmed in the studies of many ice cores and ocean sediment cores. The recent measurements of these gases show rises, which began with the widespread burning of fossil fuels in the 18th century, and has continued to rise much more sharply in the 20th and 21st centuries. Global temperatures have risen in lockstep with the rises in these greenhouse gases. Unfortunately, this already has led to increased melting of glaciers worldwide, which has caused abandonment of homes, communities, islands, peninsulas. It has also resulted in increases and broader spread of diseases, more deadly and destructive weather events, increased crop failures, human conflict (including wars and mass migrations), and eventually may lead to millions or billions of deaths, and possibly to the extinction of humans. We are at a crossroads, but some fail to recognize the signs.

The study of ice cores from Greenland and Antarctica might help to save the planet and all of its inhabitants from destruction. By this, we mean that very soon it might become difficult for humans, other primates, other animals, plants, and many other species, to survive the changes to their environments. The Earth will not disappear, and many species will be able to survive, but we may be one of the casualties of our own waste products. We cannot continue to increase in population and produce waste in the volumes that we have in the past. The study of gases in ice cores has informed us that we have not only polluted locally, but we now are polluting globally, in a demonstrably measurable way. How we deal with these issues may literally determine the fate of our species.

In addition to losing the ice itself (which helps moderate Earth's climate), some of the released microbes being might be beneficial to other organisms, and some could produce valuable products. This potential resource is being melted away by uncontrolled waste-producing activities. The search for beneficial organisms should be pursued immediately, although it should be noted that currently many pharmaceutical and drug companies are only actively advancing research on drugs and other chemicals that are of high economic value to them. Antibiotics are not among these. Research into new antibiotics has been halted by most pharmaceutical companies. However, new therapies are in development that utilize viruses, including bacteriophage, that are capable of killing pathogens. We induced bacteriophage out of a species of bacteria isolated from glacial ice. Therefore, it is possible to study these bacteria and associated viruses. There might be organisms, including bacteria and viruses, in the ice that are useful for other purposes, such as foods, food preservatives, sources of other nutrients, anti-cancer drugs, probiotics, or others.

PATHOGENS, EPIDEMICS, AND PANDEMICS

Data gleaned from the ice has informed us that enormous numbers of microbes are being released from melting ice, and that some of these are pathogens. These microbes may cause outbreaks of disease in animals, plants, or microbes, and some of the human pathogens may cause epidemics or pandemics. It may have caused disease outbreaks in the past, but this has not yet been examined. Ancient genes might enter a population of pathogens that render them more virulent, or a virus entering a bacterium could transform a harmless bacterium into a pathogenic one. This is not without precedent. Organisms, such as cholera (*Vibrio cholerae*, a bacterium) and *Microcystis aeruginosa* (a cyanobacterium) are pathogenic or toxic only because they carry genes that originated from viruses. Many strains of this cyanobacterium are harmless, but some produce liver toxins or neurotoxins. The genes for these toxins originate from a virus that infects the cyanobacteria, transforming them into toxin producers, which creates toxic algal blooms in lakes. If people or animals drink the water, they can become ill, develop liver cancers, or die. Similarly, many strains and species of *Vibrio* are harmless, but *Vibrio cholerae* carries with it a virus that has toxin genes in its genome. These toxins, once ingested by a person, cause the cells in the gut linings to break open and leak, thus leading to severe diarrhea, sometimes causing death. There may be analogous viruses in glacial ice that have never been investigated that could lead to disease in humans and other organisms.

Ice holds a large repository of viable microbes, including some pathogens that were once eradicated or otherwise went extinct. It is yet to be determined whether these organisms truly have been eradicated. For example, smallpox virus, which has killed millions of people over the centuries, was officially eradicated from the Earth in 1979. There have been no known cases of smallpox since that time. However, smallpox virus survives well when frozen. There is a small chance that somewhere in the world viable smallpox virus is frozen in glacial ice. There is a smaller risk that viable viruses would infect someone, and spread from there. It must infect host tissues very soon after it is thawed. Also, because it is a virus, while it is frozen in the ice, any damage to its genome cannot be repaired. It needs to be within a viable host cell that is still metabolically active for repair to occur. Therefore, if the virus was frozen thousands of years ago in the ice, it may not be viable. However, if it was frozen in the ice hundreds of years ago, the probability that some of the viruses are viable is higher. This implies that the return of smallpox is possible, but not probable. Attempts are currently underway to eradicate poliovirus from the world. For this virus, eradication is more likely, because it is an RNA virus. Viruses with RNA genomes experience more mutation and genetic variability, but also have no means to correct mutations. Therefore, if polioviruses are preserved in ice, their survival is more limited than for DNA viruses.

Animal vectors, such as migratory birds, mammals (including bats), and arthropods can carry pathogens from ice meltwater at high latitudes and altitudes to lower latitudes and altitudes, where the majority of the human population resides. In particular, influenza A, which is preserved in ice and mud, can be carried to populated areas, where it can be transmitted to pigs, chickens, and other animals, and that eventually lead to human infection. Recently, studies have warned that the transmission of influenza A by bats has the potential to cause epidemics, and perhaps a pandemic. This is primarily due to the virus having affinity to MHC II (major histocompatability type II) receptors, which are common to all mammals, and many other animal groups. This means that it can be more readily transferred from one species to another than can the typical strains of influenza A, which rely on attachment to sialic acid receptors on host cells. Recurring epidemics of influenza A that kill at least hundreds of thousands annually are the norm. Those that kill tens of millions of people are rare, but do occur occasionally. A flu pandemic is not only possible, but it has a high probability of occurring. The unknown is whether it will come from melting ice, or whether it will be the result of genomic mixing within a set of host organisms. Both are possible. Where this might happen is unknown, but without surveillance, its initial appearance will be missed.

MICROBIAL SURVIVAL

One thing that we have learned from studying microbes in ancient ice is that they are extremely hardy, adaptable, and resilient organisms. They likely have larger genomes, because they would need a larger repertoire of proteins and other biological molecules to adapt to the extreme conditions in the ice. But, they also need to adapt to very different conditions once they thaw from the ice. Organisms in symbiotic, mutualistic, commensal, or other close relationships with other organisms tend to have smaller genomes, because they depend on some products produced by the organisms with which they are associated. On the other hand, very adaptable organisms, one of which is *E. coli* (some of which live in your gut), have relatively larger genomes. Each environment in which they live requires a slightly different set of proteins and RNAs that allows them to obtain their nutrients, maintain an active metabolism (and other processes), excrete waste products, move around (if they are motile), and avoid being destroyed by other organisms. For example, *E. coli* normally lives in the intestines of animals, which is a dynamic anaerobic environment. It needs sets of genes that allow it to thrive there. However, it can grow aerobically as well outside of its host, and needs a different set of genes and proteins for that. Therefore, it has a larger genome that allows it to adapt to different conditions. The bacteria and fungi that we have isolated from the ice are mainly very adaptable organisms. They grow on a number of different growth media, at a wide range of temperatures, and they can grow under high pressure, as well. Initially, we thought that they would be organisms that only grew in cold places and under high pressures. It turned out that they could grow under those conditions, but they also could grow at warmer temperatures at standard atmospheric pressure. They surprised us. Prior to the study of these organisms from glacial ice, the longevity and adaptability of these organisms was unknown. Now, we can say that many survive, and they can live under a wide variety of conditions. In the future, it will be interesting to determine their genomes and the mechanisms the organisms use to survive the long periods of freezing, high pressure, and lack of light and nutrients. Some organisms produce "antifreeze proteins," and these may be common among microbes that survive in ice.

PRESERVATION OF SPECIES, AND EXTINCTIONS

As mentioned in several places in this book, a diverse assemblage of microbes exists in environmental ice worldwide. Large volumes of ice are being lost annually, some of which will never be replaced. This represents a huge number of species, some of which have simply been transported to the glacier by winds, others that live in the ice, and still others where entombment in ice is a part of their extended life cycle. Other organisms are entrapped or live in permafrost. These icy environments are vital habitats for many species of organisms. Once the ice is gone, it is likely that a large proportion of these species will go extinct. Alternatively, there are many extinct species of animals, plants, and other organisms that have been found frozen in permafrost. It may still be possible to resurrect them by extracting the DNA from dead members frozen in the ice and permafrost. The DNA sequences can be used to reconstruct the genome of these organisms to potentially produce a living member of the species once more. This has been discussed with regard to producing a living Woolly Mammoth, but it has not yet been attempted. Recently, a horse foal was recovered from 42,000-year-old permafrost. The foal still had liquid blood in its veins, which might contain intact DNA from its white blood cells. If so, this might be a candidate for resurrection. However, it might be prudent to start with a smaller organism, including unicellular, as well as multicellular, microorganisms.

SUBGLACIAL ENVIRONMENTS

The environments beneath the ice are very different from those within glacial ice, ice fields, ice domes, ice shelves, and sea ice. The basal ice of glaciers often is a mixture of ground fragments of bedrock, crushed, shredded, and damaged organisms that were once within the main body of the glacier, as well as organisms whose normal habitat is at the glacier-bedrock interface. Often, these are methanogenic Archaea, although bacterial species also are found here. From the limited data, it appears that the bases of ice fields and ice domes are similar. Ice shelves are mainly freshwater ice above, with entrapped microbes from the glacier that feeds the ice shelf, as well as accretion ice below that is primarily frozen sea water. Although there is little data regarding the microbial inclusions available, it is expected that the accretion ice contains mainly saltwater-adapted microbes, with a few microbes originating from the glacier close to the shore, and fewer of these further out on the ice shelf. As discussed in a previous chapter, sea ice contains a diverse set of microbes, as well as small and moderate-sized multicellular organisms.

The accretion ice from lakes has provided a way to study subglacial lakes without contaminating them. It has already been determined that Lake Vostok has at least two very different areas, a shallow bay that appears to have hydrothermal activity, which has a diversity of microbes, mainly bacteria, with some eukaryotes, including those that are multicellular. The other area is the main basin of the lake that contains very few organisms, with limited diversity. It shows few signs of having hydrothermal activity. Lacking an energy source would limit the possibilities for life. Lake Whillans is very different, in that it contains bacteria, which are mostly different from those found in Lake Vostok. The contents of the sediments are primarily methanogenic Archaea. It will be interesting to see the results of studies from other subglacial lakes. At this point, it appears that each lake might contain unique assemblages of microbes.

A LOOK INTO THE FUTURE

Many people assume that the ice in their drinks is safe. Most also believe that environmental ice is safe. This was on the minds of our colleagues when they were in the Caribbean having drinks with ice. Even heads of state have been presented with glacial ice. Some gladly placed the ice into their drinks, and sucked them down, while others simply looked at the ice or meltwater and passed it down the line. From what we and others have so far determined, it appears that environmental ice is not necessarily safe, and drinking water from glacial ice is a bad idea. However, it is exceedingly interesting and future studies will be valuable. Not only is it important to determine the threats from pathogens, but it will be important to determine the biodiversity of microbes in the ice and what they might indicate about ancient times on Earth. They might also produce unique and useful constituents, including antibiotics, valuable genes, and other potentially useful products. They might have systems that allow them to deal with subzero temperatures and high pressures. With the increase in the melting of ice on Earth, the release of pathogens is increasing, and the losses of potentially beneficial organisms are accelerating. There are potential losses of valuable microbes and alleles. The effects on our future are unknown, but potentially they are very significant. One possibility is that one as yet unknown pathogen may be released that might cause a serious epidemic or pandemic. Such pathogens may affect humans, but it is more likely that they might affect other species, with equally devastating results. Immediate and increased studies of microbes in environmental ice are needed.

What are the similarities and differences between life detected in ice, sediments, and permafrost? A great diversity of microbes has been found in all three matrices. Some of these organisms are capable of surviving under extreme conditions of temperature, pressure, nutrient deficiency, and oxygen concentrations. The great abundance, as well as the diversity of microorganisms, detected in these substrates has surprised microbiologists. They continue to surprise and intrigue us. Newly discovered species have been detected in all three substrates. Differences also are apparent. Archaea have been detected in subsurface habitats, including in the basal ice of glaciers, and subglacial lake sediments. To our knowledge, viruses that infect eukaryotes have not been detected in subsurface sediments, but they are present in ice. This finding likely reflects the fact that few researchers are looking for such viruses in either substrate, rather than the possibility that none exist there.

Life certainly is persistent and resilient, and under the proper preservation conditions such as in ice and other selected substrates, may be virtually immortal. Eukaryotes, archaea, and bacteria can survive for extremely long periods of time in ice. The discovery that some ice core sections contain a diverse and abundant assemblage of microorganisms while others do not, implies some, as yet unknown, relationships between the accumulation of microorganisms in time and space. Possible relationships might be to volcanic activity, climate change, change in wind patterns, meteorite impacts, disease epidemics, precipitation frequency, or a combination of these and other factors. For example, a preponderance of heat-loving microbes at some specific point in time might indicate fallout from a volcanic eruption, as we found with some of our studies. The presence of temperate, tropical, and heat-loving microbes in the ice core record suggests several possibilities. The climate of the polar regions has been much warmer than today, which permitted the local growth of some microorganisms. Or these microorganisms might have been blown onto the polar ice caps when changing wind patterns permitted their spread from more temperate or tropical latitudes, especially during volcanic eruptions.

Many microorganisms including viruses can be airborne for long periods of time and over long distances while remaining viable, or infectious if they are pathogens. Thus, the patterns of distribution of microorganisms within the ice core record may be a useful way to corroborate changing climate conditions as determined by other methods. Decreased global temperatures are associated with increased aridity, wind-blown dust, and microorganisms that accumulate in ice cores. Unfortunately, we do not yet know if specific microorganisms are associated with specific types of particles from specific sources. Much more research is needed in this area. Similarly, the detection of human, animal, or plant pathogens might point to a disease epidemic or pandemic. Many pathogens are airborne and/or waterborne (e.g., the bacterium that causes cholera, influenza A virus, poliovirus, marine caliciviruses, tobamoviruses, and many other pathogens). It may be possible to track the historical time course of some human, animal, or plant diseases by examining the microbial ice record. The prevalence in ice of microbes that prefer temperate or tropical climates suggests long-range transport of microbes in the atmosphere and in the oceans, as well as shifting wind patterns and local climatic changes.

These findings lend some support to the Panspermia and Exogenesis hypotheses. Both assume that microbial communities can retain viability during long-term migrations perhaps even to and/or from other planets. The implications are that terrestrial life may have originated from extraterrestrial sources, or alternatively that life originated on Earth and has migrated to other worlds. Reports that certain microorganisms can survive the temperatures and pressures encountered during reentry into Earth's atmosphere from outer space also support the plausibility of such theories. Eukaryotes and bacteria are known to be capable of long-term anabiosis (i.e., dormancy). It may now be possible to study mechanisms of cryopreservation of more complex eukaryotic organisms using yeasts and filamentous fungi isolated from ice as models. A multitude of organisms and applications remain to be discovered. But, the evidence is overwhelming that environmental ice contains large numbers of living and dead microbes, representing millions of years of microbial history. Study of these ice samples has unlocked a huge vault of information about Earth's history, and more is being learned every day due to the rapid melting of ice on Earth. Potentially dangerous microbes are being released in ever increasing numbers. It will be interesting to watch this "Pandora's box" as it opens wider, and faster. The grand experiment is underway, and we are in it.

SOURCES AND ADDITIONAL READINGS

Augustin, L., C. Barbante, P.R. Barnes, et al. (55 authors). 2004. Eight glacial cycles from an Antarctic ice core. *Nature* 429: 623-628.
Castello, J.D., and S.O. Rogers. 2005. Life in Ancient Ice. Princeton, NJ: Princeton University Press.
Cheng, J., J. Abraham, Z. Hausfather, and K.E. Trenberth. 2019. How fast are the oceans warming? *Science* 363 128-129.
Christner, B.C., J.C. Priscu, A.M. Achberger, C. Barbante, S.P. Carter, K. Christianson, A.B. Michaud, J.A. Mikucki, A.C. Mitchell, M.L. Skidmore, T.J. Vick-Majors, and the WISSARD Science Team. 2014. A microbial ecosystem beneath the West Antarctic ice sheet. *Nature* 512: 310-313.
Rogers, S. O. 2017. Integrated Molecular Evolution, 2nd ed. Boca Raton, FL: Taylor & Francis Group.
Rogers, S.O, W.T. Starmer, and J.D. Castello. 2004. Recycling of pathogenic microbes through survival in ice. *Med. Hypoth.* 63: 773-777.
Rogers, S.O., Y.M. Shtarkman, Z.A. Koçer, R. Edgar, R. Veerapaneni, and T. D'Elia. 2013. Ecology of subglacial Lake Vostok (Antarctica), based on metagenomic/meta-transcriptomic analyses of accretion ice. *Biology* 2: 629-650.
Shtarkman Y.M., Z.A. Koçer, R. Edgar R, R.S. Veerapaneni, T. D'Elia, P.F. Morris, and S.O. Rogers. 2013. Subglacial Lake Vostok (Antarctica) accretion ice contains a diverse set of sequences from aquatic, marine and sediment-inhabiting Bacteria and Eukarya. PLoS ONE 8(7): e67221. doi:10.1371/journal.pone.0067221.
Shoham, D., A. Jahangir, S. Ruenphet, and K. Takehara. 2012. Persistence of avian influenza viruses in various artificially frozen environmental water types. Influenza Res. Treat. dx.doi.org/10.1155/2012/912326.
Smith, A.W., D.E. Skilling, J.D. Castello, and S.O. Rogers. 2004. Ice as a reservoir for pathogenic animal viruses. *Med. Hypoth.*, 63: 560-566.
Zhang, G., D. Shoham, D. Gilichinsky, S. Davydov, J.D. Castello, and S.O. Rogers. 2006. Evidence for influenza A virus RNA in Siberian lake ice. *J. Virol.*, 80: 12229-12235.

LINKS

https://en.wikipedia.org/wiki/Bird_migration
https://en.wikipedia.org/wiki/Influenza_A_virus
https://en.wikipedia.org/wiki/Spanish_flu
https://en.wikipedia.org/wiki/Woolly_mammoth
https://www.livescience.com/65268-oldest-liquid-blood-siberian-foal.html
https://www.sciencedaily.com/releases/2019/02/190220133534.htm
https://www.sciencemag.org/news/2014/10/virus-resurrected-700-year-old-caribou-dung

Index

Note: Numbers in *italic* indicate a figure and page numbers in **bold** indicate a table on the corresponding page.

Printed in the United States
by Baker & Taylor Publisher Services